Springer Tracts in Mechanical Engineering

Springer Tracts in Mechanical Engineering (STME) publishes the latest developments in Mechanical Engineering—quickly, informally and with high quality. The intent is to cover all the main branches of mechanical engineering, both theoretical and applied, including:

- Engineering Design
- Machinery and Machine Elements
- Mechanical structures and Stress Analysis
- Automotive Engineering
- Engine Technology
- Aerospace Technology and Astronautics
- Nanotechnology and Microengineering
- Control, Robotics, Mechatronics
- MEMS
- Theoretical and Applied Mechanics
- Dynamical Systems, Control
- Fluids mechanics
- Engineering Thermodynamics, Heat and Mass Transfer
- Manufacturing
- Precision engineering, Instrumentation, Measurement
- Materials Engineering
- Tribology and surface technology

Within the scopes of the series are monographs, professional books or graduate textbooks, edited volumes as well as outstanding PhD theses and books purposely devoted to support education in mechanical engineering at graduate and post-graduate levels.

Indexed by SCOPUS and Springerlink. The books of the series are submitted for indexing to Web of Science.

To submit a proposal or request further information, please contact: Dr. Leontina Di Cecco Leontina.dicecco@springer.com or Li Shen Li.shen@springer.com.

Please check our Lecture Notes in Mechanical Engineering at http://www.springer.com/series/11236 if you are interested in conference proceedings. To submit a proposal, please contact Leontina.dicecco@springer.com and Li.shen@springer.com.

More information about this series at http://www.springer.com/series/11693

Iulian Popescu · Liliana Luca ·
Mirela Cherciu · Dan B. Marghitu

Mechanisms for Generating Mathematical Curves

Iulian Popescu
Calea Bucureşti
Facultatea de Mecanică
Craiova, Romania

Liliana Luca
Calea Eroilor
Universitatea Constantin Brâncuşi
Târgu Jiu, Romania

Mirela Cherciu
Calea Bucureşti
Facultatea de Mecanică
Craiova, Romania

Dan B. Marghitu
Mechanical Engineering Department
Auburn University
Auburn, AL, USA

ISSN 2195-9862 ISSN 2195-9870 (electronic)
Springer Tracts in Mechanical Engineering
ISBN 978-3-030-42170-0 ISBN 978-3-030-42168-7 (eBook)
https://doi.org/10.1007/978-3-030-42168-7

This Springer imprint is published by the registered company Springer Nature Switzerland AG
The registered company address is: Gewerbestrasse 11, 6330 Cham, Switzerland

Foreword

As part of the realm of mechanical engineering, one of the oldest and broadest of the engineering disciplines, the Theory of Machines and Mechanisms is one of the most challenging but, at the same time, a very exciting field of innovation and research. Starting with the simple mechanisms invented and used by ancient Egyptians until nowadays, discovering new mechanical systems to transmit and transform motion and power has been part of humanity history. By way of example only, some of the well-known engineers of all time that contributed to the field of Machine and Mechanisms include Archimedes, Leonardo da Vinci, George Stephenson, among others.

Following the great examples from the predecessors, the authors of this book propose new and original mechanisms capable to generate complex mathematical curves that could be used in modern machine design. The purpose of this book is to present the reader with a comprehensive text that includes theory and examples. Useful and innovative analytical techniques provide the reader with powerful tools for mechanisms synthesis and analysis necessary for mechanical design. Furthermore, this book may also serve as a reference for mechanical designers and as sourcebook for researchers.

Iulian Iordăchiţă
Research Professor
Johns Hopkins University
Baltimore, MD, USA

Preface

To write a book, today, about mechanisms for generating mathematical curves, is an act of bravery! Nowadays, when the software packages can calculate mechanisms and the computer numerical control machines (CNC) can generate or create any curve, it seems there is no place for mechanisms generating mathematical curves. However, these mechanisms can be used in various automatic packaging systems, some agricultural machines, in textile machines, in small workshops, for drawing or welding some caps tanks and in toys manufacturing.

Edmund Hillary, who conquered Everest in 1953 together with Tenzing Norgay, was asked: “Why so much effort for a mountain top?”, and he said: “Because it exists!”

Something like this is, also, in the case of mechanisms that generate mathematical curves: *they do exist*!

And because they exist, they have to be calculated, which is not easy because they are quite complicated.

Therefore, the book starts with mathematical considerations. The synthesis of new, original mechanisms that can generate these curves is performed, and then, their kinematics is studied to confirm the drawing of those curves. The synthesis is made with creativity, which a computer is not able to do. The analysis uses the closed-loop method and in some cases the distances method. We give notions of structure and kinematics strictly necessary to calculate these mechanisms.

For each mechanism, a computer program was developed, so that the desired curves, the successive positions of the mechanisms, and also the kinematic diagrams were obtained.

We looked for other kinematic possibilities of creating mechanisms, as well.

The book can also be seen as a collection of mechanism exercises, useful to those who want to learn mechanism kinematics (students), and those engineers who have studied this subject once and that now work with design calculations for mechanisms.

The equations given for the points that draw the mathematical curves can be used to create programs for the CNC machine tools that can generate these curves.

This book is the result of many years of research by the authors; these studies were published in various books and articles in Romania and are cited throughout this work, with the intention of becoming known also to English readers.

We hope this book will be useful!

Craiova, Romania Iulian Popescu
Târgu Jiu, Romania Liliana Luca
Craiova, Romania Mirela Cherciu
Auburn, USA Dan B. Marghitu

Contents

About the Authors

Emeritus Professor Dr. Eng. Iulian Popescu, full member of the Academy of Technical Sciences of Romania, taught Mechanisms and Machines Theory at the Faculty of Mechanics, University of Craiova, for 35 years, guided 25 Ph.D. students in the field of mechanisms, and printed 31 books of mechanisms in Romanian language. The main titles of the mechanisms books written by the authors of this work are given below.

1. Popescu I.—The Theory of Mechanisms and Machines. SITECH Publishing House, Craiova, 1997.
2. Popescu I.—Matrix analysis mechanisms. Reprography of the University of Craiova, 1977.
3. Popescu I.—Mechanisms. New algorithms and programs. Reprography of the University of Craiova, 1997.
4. Popescu I., Luca L., Cherciu M.—Trajectories and movement laws of some mechanisms, Sitech Publishing House, Craiova, 2011.
5. Popescu I., Luca L., Mitsi Sevasti—Geometry, structure and kinematics of some mechanisms, Sitech Publishing House, Craiova, 2011.
6. Popescu I., Mîlcomete D.—Researches on synthesis and optimization of mechanisms, Sitech Publishing House, Craiova, 2006.
7. Popescu I., Sass, L.—Mechanisms generating curves, Ed. "Scrisul Românesc", Craiova, 2001.
8. Popescu I., Ungureanu, A.—Structural and Cinematic Synthesis of Bar Mechanisms, "Universitaria" Publishing House Craiova, 2000.

9. Popescu, I.—Curves and aesthetic surfaces: geometry, generation, applications. Sitech Publishing House, Craiova, 2013.
10. Popescu, I.—Structure and kinematics of some mechanisms generating curves and aesthetic surfaces, SITECH Publishing House, Craiova, 2012.
11. Popescu, I., Luca L., Cherciu M.—Structure and kinematics of mechanisms. Applications. Sitech Publishing House, Craiova, 2013.

Liliana Luca holds Ph.D. Professor at the Faculty of Engineering, Constantin Brancuşi University of Târgu Jiu, Romania.

- Her Ph.D. thesis is entitled "Contributions to The Synthesis of Some Prehensile Mechanisms," based on biomechanisms, and she obtained her Ph.D. in 2000 at the University of Craiova, Romania.
- She is the first author or coauthor of six mechanisms books published in Romania.
- She has published more than 70 articles in ISI/BDI indexed journals and more than 60 articles in proceedings of national and international symposiums.
- She received the Best Paper Award (2012) for "*Studies regarding generation of aesthetics surfaces with mechanisms*," Proceedings of the 3rd International Conference on Design and Product Development, Montreux, Switzerland, ISBN 978-1-61804-148-7.

Mirela Cherciu is Associate Professor at the Faculty of Mechanics, University of Craiova, since 1988, and holds Ph.D. in mechanical engineering. She has some contributions in the field of mechanism's precision, working under the guidance of Emeritus Professor Dr. Eng. Iulian Popescu, her Ph.D. supervisor. The results of her research could be found in the Ph.D. thesis, completed in 2000, "Contributions to the study of the influence of tolerances on the precision of the mechanisms" and in a number of five books published in Romania and over 70 scientific papers published in national and international conferences.

Dan B. Marghitu is Professor of mechanical engineering in the College of Engineering at Auburn University, USA. He holds Ph.D. from Southern Methodist University, Dallas, Texas, and a DEA from Paul Sabatier University, Toulouse, France. He has published over 70 research papers in journals and has authored or coauthored six textbooks in the areas of dynamics, robotics and mechanisms. He has done work in impact dynamics with applications to robotics systems and nonlinear dynamics with application to human and animal locomotion. He has been PI/Co-PI of funded projects and has organized and chaired many international conferences.

Part I
Basic Theory of Mechanisms

Chapter 1
Structural Analysis of Planar Mechanisms

Abstract The basic elements of the planar mechanism's structure are given: kinematic joints, symbols, kinematic links, degrees of freedom, structural substitution of the higher pairs with contact by line or point, Assur's principle, particular cases of dyads, decomposition in kinematic groups, types of triads, and the transition from kinematic diagrams to structural diagrams.

1.1 Kinematic Joints

The analysis of the mechanisms studies the existing mechanisms, in order to know their capabilities and improve or adapt them to other functional requirements.

Currently, there is a wide range of mechanisms used in different areas. To study any mechanism, the calculation methods must be as general as possible.

Existing mechanisms are not completely different, and they have in common a number of features, which are emphasized by structural analysis [2].

The elements of mechanisms are considered to be rigid and non-deformable. Theory of Mechanisms is trying to establish calculation methods for categories of mechanisms [6], but there are mechanisms empirically produced, which can not fit easily into these categories.

For mechanisms, the bodies, called mechanism elements, are linked together. These connections restrain the possibilities of motion.

The connection of two elements, that applies constraints on their relative movement, is called *kinematic pair* or *joint* [3, 6].

Figure 1.1 shows an element that moves in xy plane. Any movement can be divided into two translations along the axes of the system, $\overline{V_x}$ and $\overline{V_y}$, and a rotation around z-axis, perpendicular to xy plane. However, if the kinematic element 1 is linked with the element 2, then this connection (Fig. 1.2) makes the two components not to move independently, and it is a kinematic pair.

Another representation of the kinematic pair is shown in Fig. 1.3. Link 1 has a complete rotation (crank), and it is attached to a frame called *base or ground*, which is a fixed element. A free rigid body in planar motion has three possible motions, and in this case, the crank has just a single motion, a rotation about y-axis. Since a free

I. Popescu et al., *Mechanisms for Generating Mathematical Curves*,
Springer Tracts in Mechanical Engineering,
https://doi.org/10.1007/978-3-030-42168-7_1

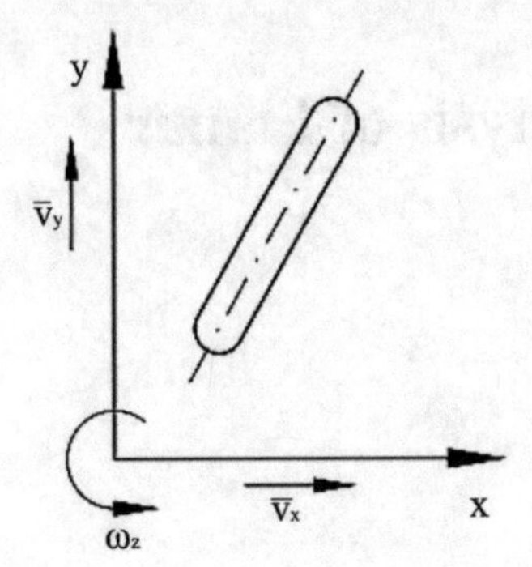

Fig. 1.1 Kinematic element

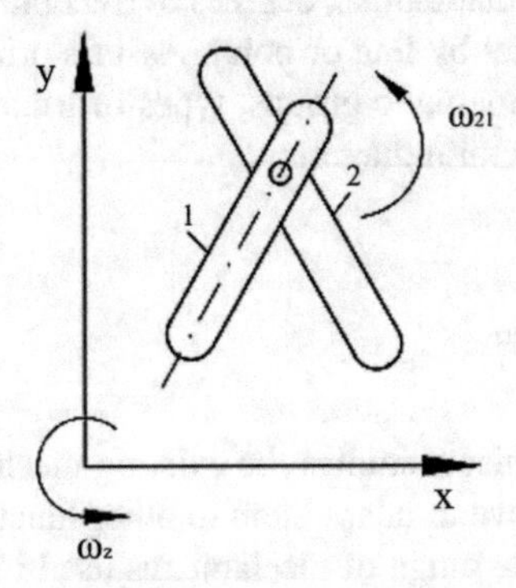

Fig. 1.2 Kinematic pair

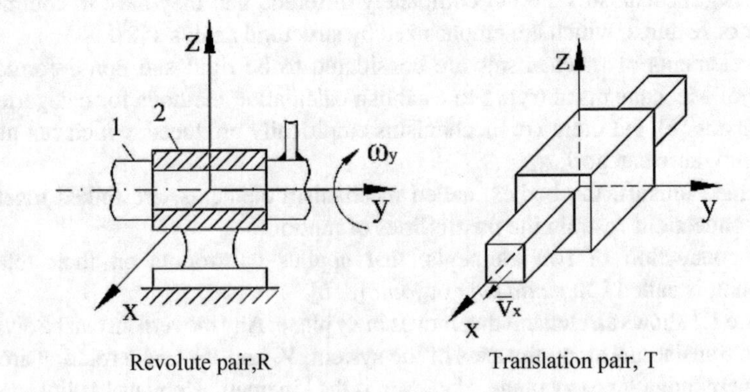

Fig. 1.3 Fifth class pairs

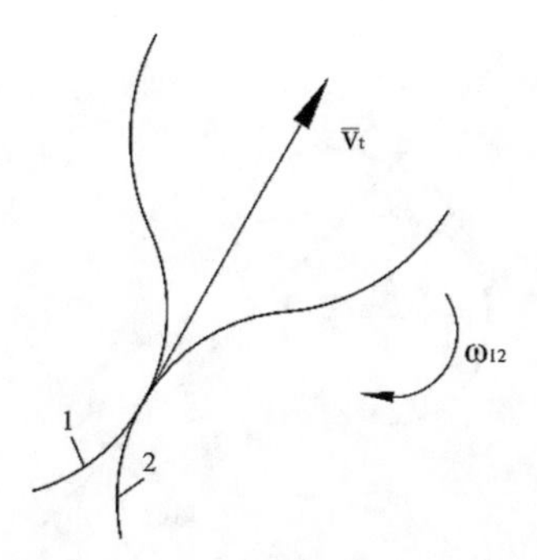

Fig. 1.4 Fourth class higher pair

body in space has six possible movements, and the pair in Fig. 1.3 allows a single relative movement between elements, it proves that this connection suppresses five movements.

This kinematic pair is called a *fifth class revolute pair (or revolute joint)* and is symbolized by R.

In Fig. 1.3, there are also two elements which allow just a relative translation between the links, which consequently, is a *fifth class translation pair.*

This connection is named *prismatic pair or slider*, and it is symbolized by T (translation) or by P (prismatic).

If the contact between the two elements of the pairs is a surface-contact, as shown in Fig. 1.3, the pair is called a *lower pair*.

For planar mechanisms, there is one more category of pairs, shown in Fig. 1.4. In this case, the contact between the two elements is a point-contact (or line-contact if taking into account the thickness of the elements). The relative motion of the elements can be a rolling movement with ω_{12} the relative angular velocity, so that this is a *higher fifth class pair.* When there are both rolling and sliding motions in the direction of the common tangent at the contact point, $\overline{V_t}$, we have a *higher fourth class pair*. This kinematic pair allows two movements and prevents four motions.

These are the kinematic pairs common to planar mechanisms.

The lower pairs are reversible, i.e., the motion is the same no matter if element 1 rotates (or translates) relative to element 2 or 2 rotates (or translates) relative to 1 (Fig. 1.3).

In the case of higher pairs, the motions are different, for example, in Fig. 1.5, if wheel 1 rolls without sliding over the rail 2 (higher fifth class pair), a point on the wheel will generate an orthocycloid. If the rail is rotating on the wheel (Fig. 1.6),

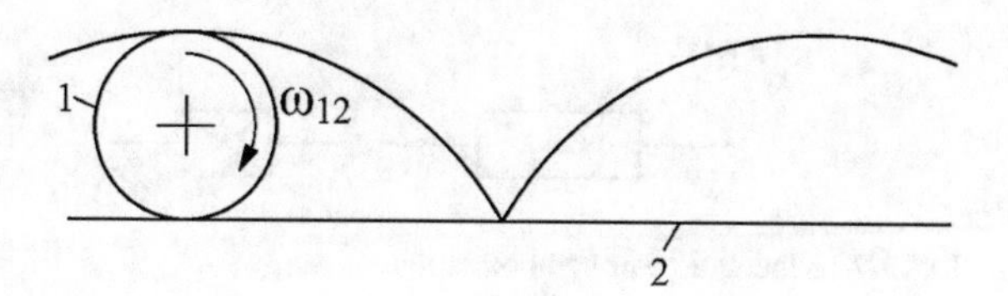

Fig. 1.5 Generation of the orthocycloid

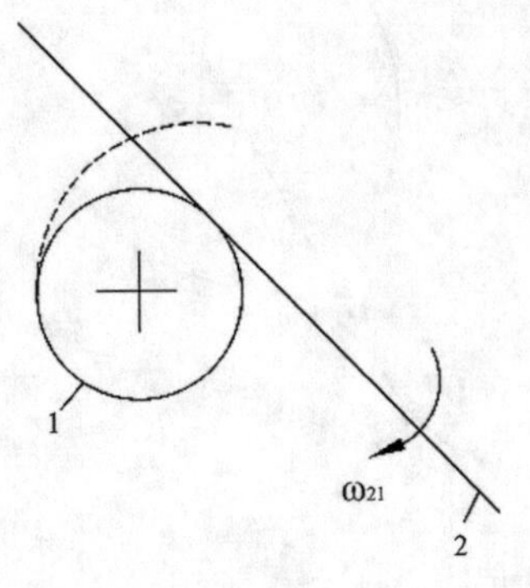

Fig. 1.6 Generation of the involute

it results in a different curve, orthogonal to all the tangents to the circle, named an involute. Therefore, higher pairs are irreversible.

The kinematic pair representation is in Fig. 1.7:

(a) fifth class revolute pair (joint);
(b) fifth class revolute pair with a fixed element called base or frame;
(c) fifth class translation pair or slider;
(d) fifth class translation pair or slider, a particular case with the length of the element l equal to zero;
(e) higher pair.

It is specified that technologically it is easier to manufacture a cylindrical pair than a prismatic one. Therefore, for the planar mechanisms instead of the slider, usually, a

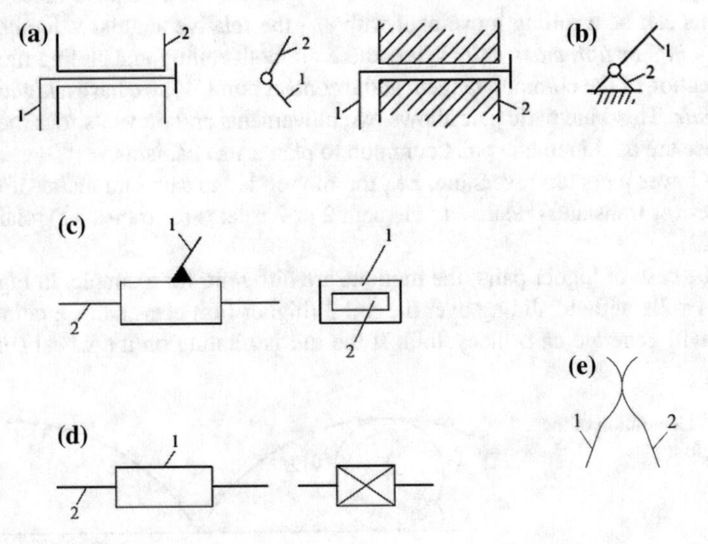

Fig. 1.7 Kinematic pair symbolization

cylindrical pair is used (indicated in Fig. 1.8). The revolute motion is constrained by the rest of the kinematic chain. So, it becomes a slider, but with a different symbol.

In the case of spatial mechanisms, including the industrial robots, the slider is drawn as a rectangle with both diagonals (Fig. 1.7d).

The symbols used for the kinematic elements are shown in Fig. 1.9:

(a) element that is part of two revolute pairs;
(b) element that is part of a revolute pair and a translation pair, guide 2 is an element that does not matter in this case, only as direction;
(c) element that is part of a revolute pair and a translation pair, with zero length (element 1);
(d) element that is part of two translation pairs;
(e) element that is part of three revolute pairs with the axes in the same plane;

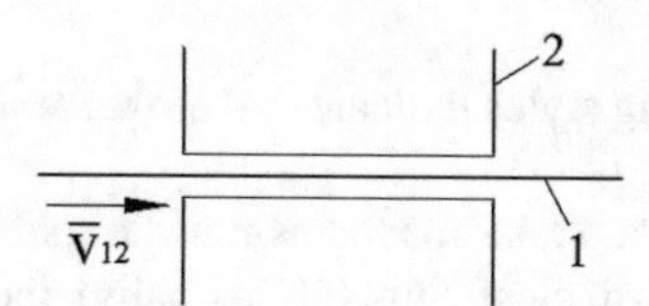

Fig. 1.8 Cylindrical pair transformed in a plane prismatic pair

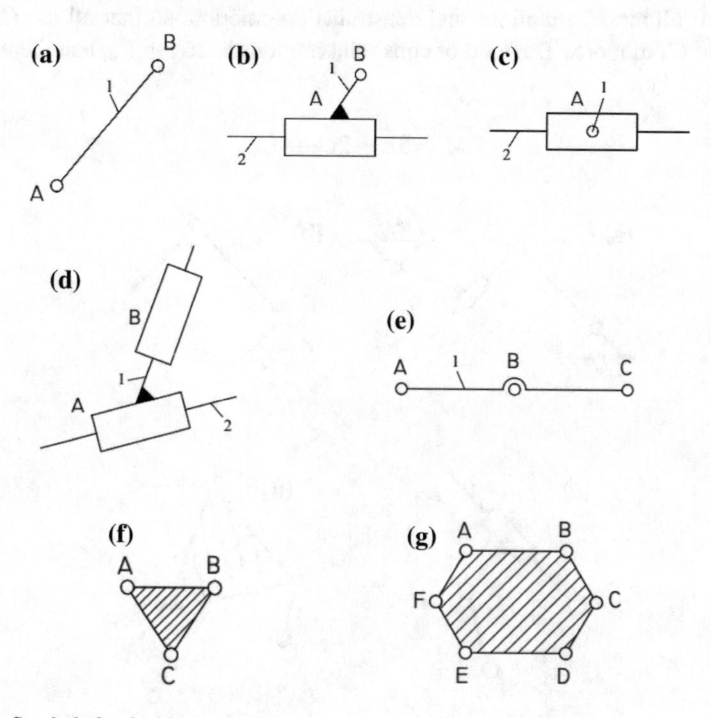

Fig. 1.9 Symbols for the kinematic elements

(f) element that is part of three revolute pairs with the axes in different planes;
(g) polygonal element (plate type) that is part of a large number of joints.

1.2 Kinematic Links and the Degree of Freedom

A sequence of elements linked by kinematic pairs is called *kinematic chain* [3, 6]. Examples are given in Fig. 1.10:

(a) RRPR;
(b) RRR;
(c) RPRR;
(d) RRRRRRR.

In the drawings, hatching styles indicate that *those elements are linked to the base* (frame).

If a kinematic chain or a planar mechanism has n mobile elements linked by C_5 fifth class pairs and C_4 fourth class pairs (higher pairs), then from all the $3n$ possible motions, the pairs constrain some motions: A fifth class pair allows a single motion, so it constrains two motions, and therefore, C_5 pairs constrain $2C_5$ motions. A fourth class pair allows two motions and constrains one motion, so that all the C_4 pairs constrain C_4 motions. The total of constraint motions is $2C_5 + C_4$, and therefore, it remains (1.1):

$$M = 3n - 2C_5 - C_4 \tag{1.1}$$

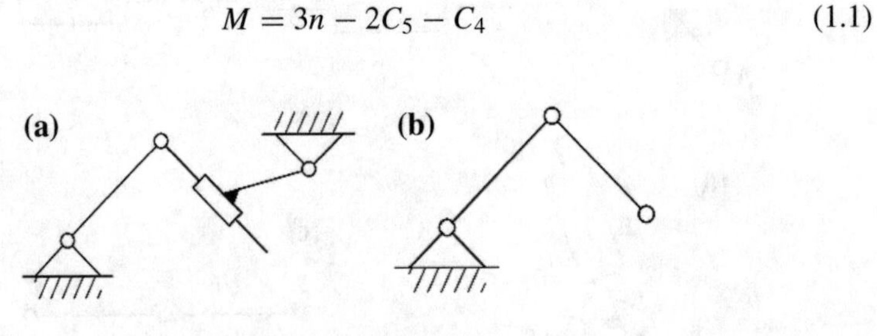

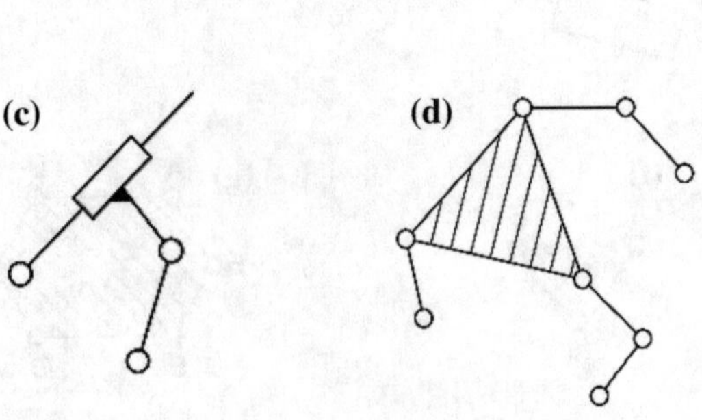

Fig. 1.10 Kinematic links

M is called the mobility of the system, and it is equal to the number of *degrees of freedom (DOF)* [1, 3]. The number of DOF is *equal to the number of driving links* of the mechanism. These mechanisms are *third family* mechanisms. The *third family* mechanisms have three motions.

A kinematic chain with movements defined in relation to the base is called *mechanism.*

There are also planar mechanisms of the *fourth family*, that means they lost four motions of the elements from all the six possible movements. This is the case of the mechanisms that have only translation pairs, as in Fig. 1.11. The mechanism has only translation pairs at A, B, and C. The motion at B is a translation in a direction that is not parallel to the axes and can be divided into two translational components along the axes. For a chain with n mobile elements, there are $2n$ possible motions in the plane, and each fifth class pair have just one motion, so it looses one of two possible motions, namely the chain loses C_5 motions, remaining (1.2):

$$M = 2n - C_5 \tag{1.2}$$

In the case shown in Fig. 1.11, it results $M = 2(2) - 3 = 1$, i.e., if the link 1 is driven by the force P, which provides the velocity $\overline{V_1}$, then the link 2 will move up driven by the force Q.

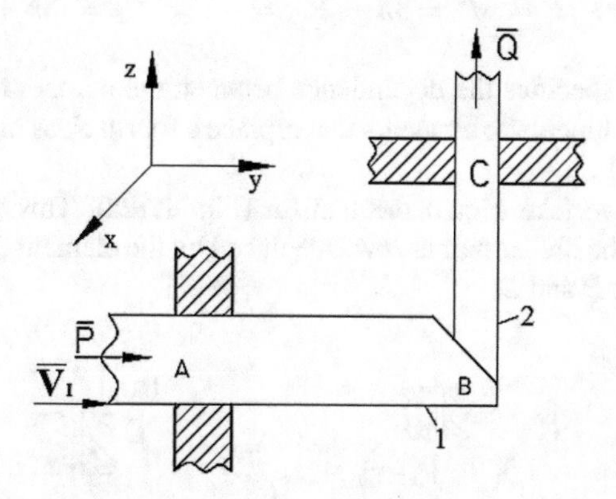

Fig. 1.11 Mechanism of the fourth family

1.3 Substitution of Fourth Class Joints

For some kinematic calculations, it is more convenient not to work with higher pairs, therefore, fictional, *these pairs are substituted by replacement kinematic chains*, obtaining replacement mechanisms (only for calculation, actually mechanisms are keeping their higher pairs) [3, 4]. It is demonstrated that a fourth class higher pair can be replaced, for the planar mechanisms, with a chain made up of elements and fifth class pairs, if they meet the following conditions:

(a) The degree of freedom for the replacer chain must be equal with the freedom degree of the replaced pair.
(b) The instantaneous relative motion between the elements which form the higher pair has to remain the same.

The number of degrees of freedom for the replacer chain is:

$$M' = 3n - 2C_5 \tag{1.3}$$

A kinematic pair that links two elements constrains some of their possible motions. A fourth class higher pair introduces only one linking condition in plane, and therefore, it does not offer any extra degrees of freedom. Actually, it constrains one degree of freedom, that means it can be considered having the degrees of freedom:

$$M'' = -1 \Rightarrow M' = M'' = 3n - 2C_5 = -1 \Rightarrow C_5 = (3n + 1)/2 \tag{1.4}$$

Equation (1.4) specifies the dependence between the number of fifth class pairs and the number of kinematic elements that replace a fourth class higher pair (usually $n = 1$ and $C_5 = 2$).

For example, we take a cam mechanism (Fig. 1.12a). This mechanism has a higher pair at B. The higher pair at B was replaced by the element 3 and two revolute kinematic pairs, at B and D.

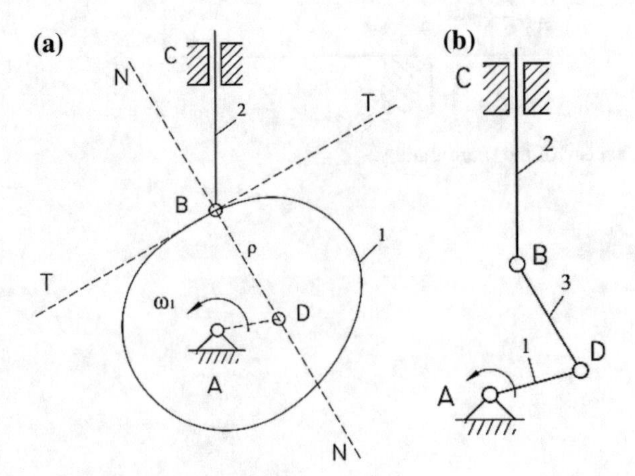

Fig. 1.12 Structural replacement of the higher pairs

The length of the element 3, BD, is equal to the curvature radius ρ of the cam shape at B (Fig. 1.12a) to comply with the condition (Fig. 1.12b).

The radius of curvature is located on the normal NN, perpendicular to the tangent in the contact point, TT. The new element BD is parallel to NN. Now, it will work with the mechanism shown in Fig. 1.12b, which has only fifth class pairs. It is stated that for each position of the mechanism shown in Fig. 1.12a, there are other dimensions of the mechanism from Fig. 1.12b, by modifying the radius and the center of curvature.

1.4 Assur's Principle and Kinematic Groups

The Assur's principle shows that any planar mechanism is made of one or more driving links, a base, and one or more kinematic chains with zero degree of freedom. The kinematic chains with zero DOF are called *kinematic groups* [3, 4].

If a kinematic group has its extreme pairs connected to the base, it becomes a rigid.

The degrees of freedom of the planar chains are:

$$M = 3n - 2C_5 \Rightarrow M = 3n - 2C_5 = 0 \Rightarrow C_5 = 3n/2 \tag{1.5}$$

Equation (1.5) indicates the connection between n and C_5, which forms different kinematic groups. The common groups will be those in Table 1.1.

A first kinematic group consists of two elements and three fifth class joints, called *second class group*, or *dyad*. In Fig. 1.13, it is represented a dyad having only revolute joints or RRR, that is called *dyad of first configuration* [3]. This dyad is of second class because it has two free joints. The class of the group is given by the maximum number of joints that forms a loop (in this case, a loop is considered BC or CD). The order of the group is given by the number of the extreme joints that link the group to the mechanism.

Table 1.1 Kinematic groups

n	2	4	6	8	10	...
C_5	3	6	9	12	15	...

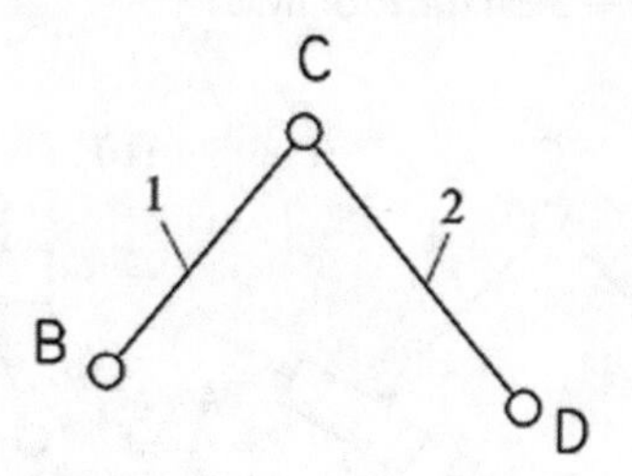

Fig. 1.13 RRR dyad

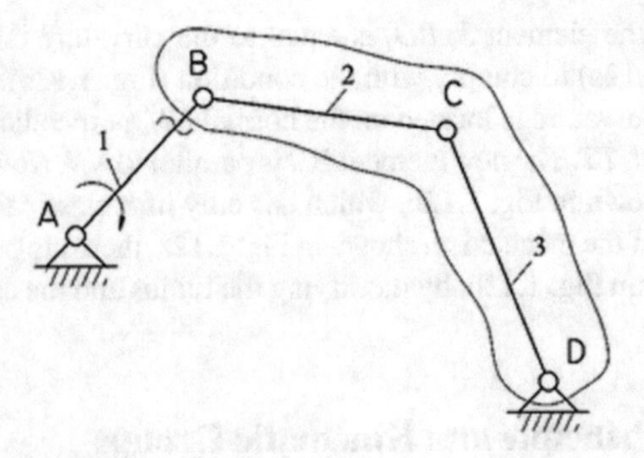

Fig. 1.14 Mechanism with one RRR dyad

For example, in Fig. 1.14 it is given a four-bar linkage, consisting of a driving link, the dyad *BCD* and the base *AD*. The element that performs full rotations is called crank. The element in plane motion is called connecting rod. The element with a limited rotating oscillatory motion is called rocker.

The mechanism in Fig. 1.15 has the degree of freedom $M = 3(6) - 2(8) = 2$, so it has two driving links and two RRR dyads.

Another dyad, with an extreme prismatic joint is shown in Fig. 1.16a. The dyad is RRP (PRR) type, that is called *dyad of second configuration*. The guide 4 is not part of the dyad; it is a different element (base, driving link or belonging to other kinematic groups); it is mentioned here only as a direction of translation. If the length

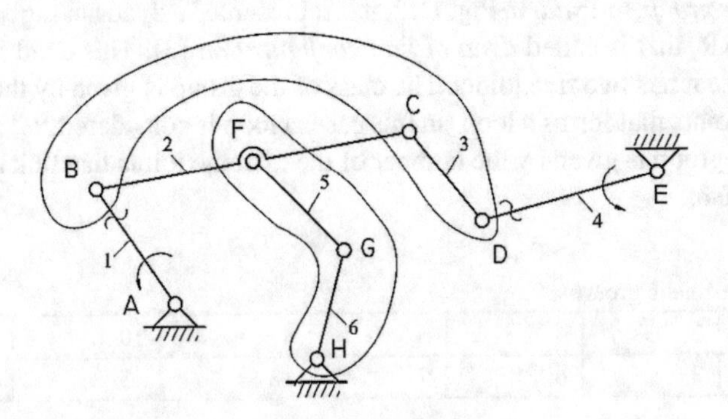

Fig. 1.15 Mechanism with $M = 2$ and two RRR dyads

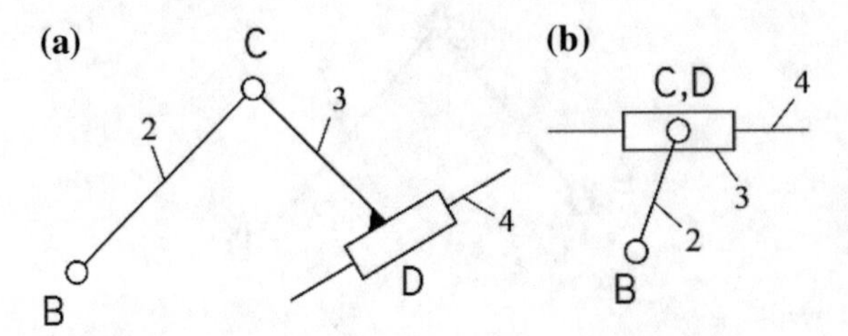

Fig. 1.16 RRP dyad

of the element BC becomes zero, we obtain a particular case of the dyad, shown in Fig. 1.16b, which is a part, for instance, of the slider-crank mechanism, as shown in Fig. 1.17.

In the case of an interior prismatic joint, it is obtained a *dyad of third configuration* (RPR), in Fig. 1.18a, with the particular case $l_3 = 0$ in Fig. 1.18b. This is the case for the crank-and-rocker mechanism in Fig. 1.19.

If two extreme prismatic joints are used, as in Fig. 1.20, a *dyad of fourth configuration*, PRP type is obtained, with the particular cases in Fig. 1.21:

- with $l_2 = 0$, Fig. 1.21a;
- with $l_3 = 0$, Fig. 1.21b;
- with $l_2 = l_3 = 0$, Fig. 1.21c.

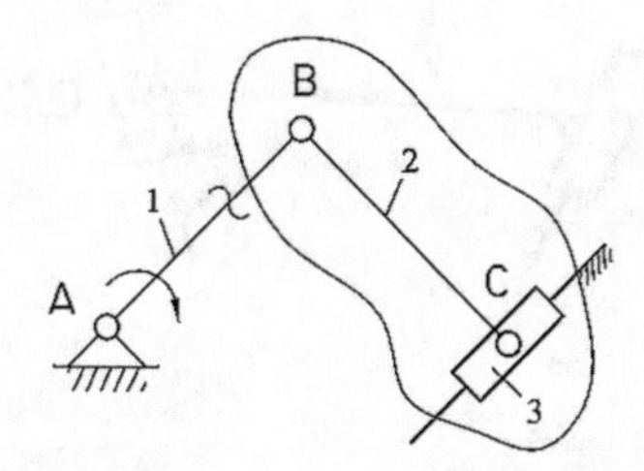

Fig. 1.17 Mechanism with RRP dyad

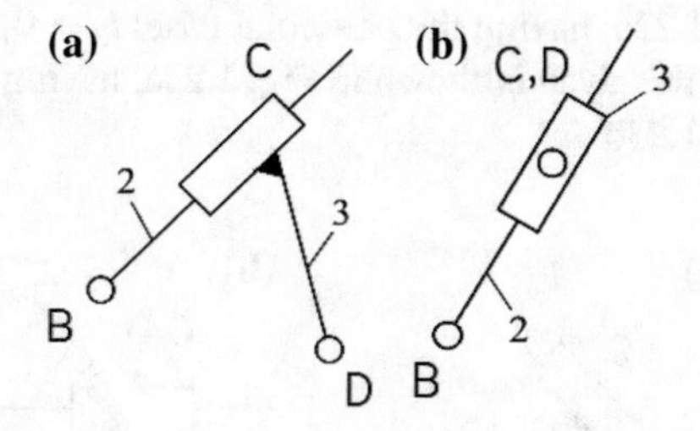

Fig. 1.18 RPR dyad

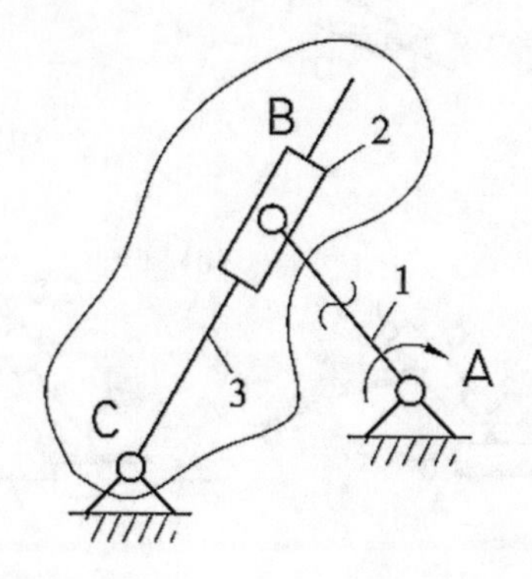

Fig. 1.19 Mechanism with RPR dyad

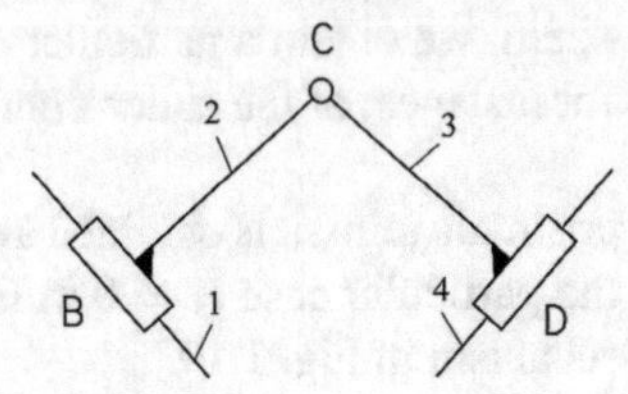

Fig. 1.20 PRP dyad

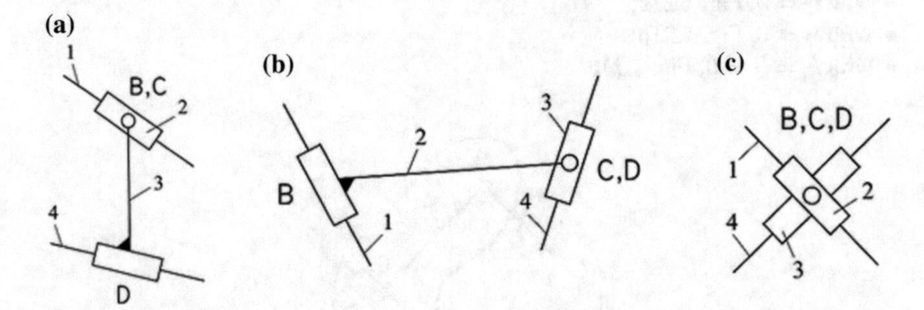

Fig. 1.21 PRP dyad, particular cases

In the case of a single extreme revolute joint, it is given a *dyad of fifth configuration* (RPP or PPR) in Fig. 1.22a, having the particular case: $l_3 = 0$, in Fig. 1.22b.

Another version of this dyad is shown in Fig. 1.23a, having the particular case $l_3 = 0$, as shown in Fig. 1.23b.

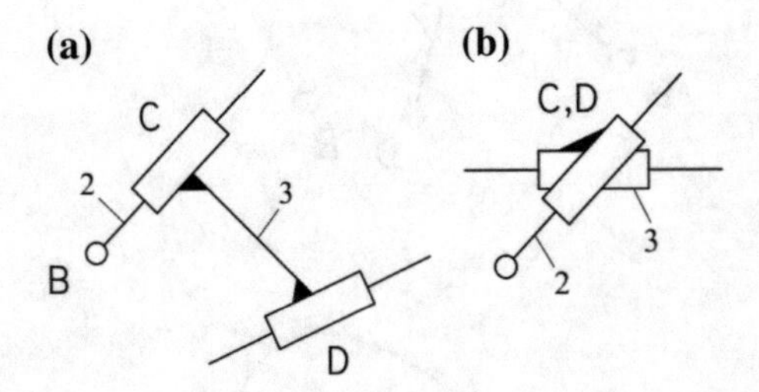

Fig. 1.22 RPP dyad

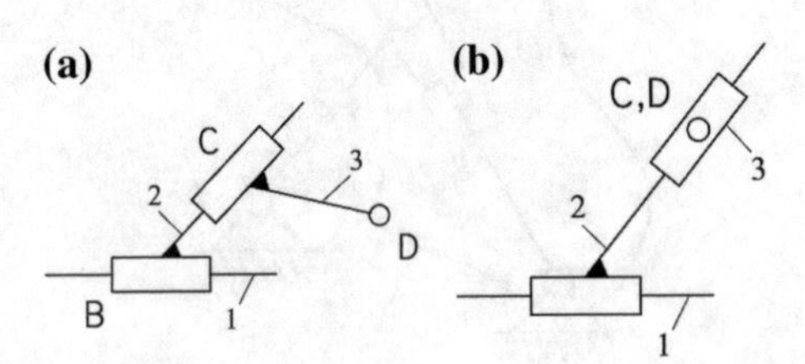

Fig. 1.23 PPR dyad

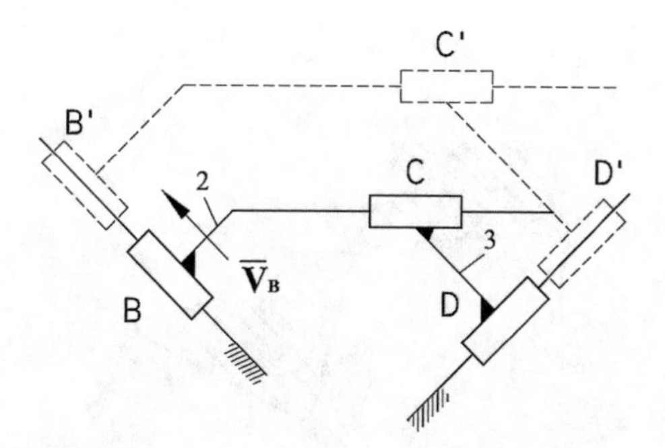

Fig. 1.24 Mechanism with plane translations (PPP)

The version PPP in Fig. 1.24 is not a dyad, since the extreme joints being connected to the base allow the motion. It results a mechanism which allows only plane translations, so $M = 2n - C_5 = 2.2 - 3 = 1$.

In Table 1.1, there are given other kinematic groups:

- For $n = 4$ and $C_5 = 6$, it results the group of *third class* or *triad* (the maximum loop given by the joints is 3), of the third order (Fig. 1.25);
- For $n = 6$ and $C_5 = 9$, it results the triad of the fourth order (Fig. 1.26);
- For $n = 8$ and $C_5 = 12$, it results the triad of the fifth order (Fig. 1.27);
- For $n = 4$ and $C_5 = 6$, it results the tetrad of the second order (Fig. 1.28).

In a similar manner, other kinematic groups can be built, but they are rarely used because they have more elements, causing large errors in the achieved laws of motion. The class of a mechanism is given by the highest class of a group from its structure. Therefore, any planar mechanism will fit in a class and will be calculated with specific methods.

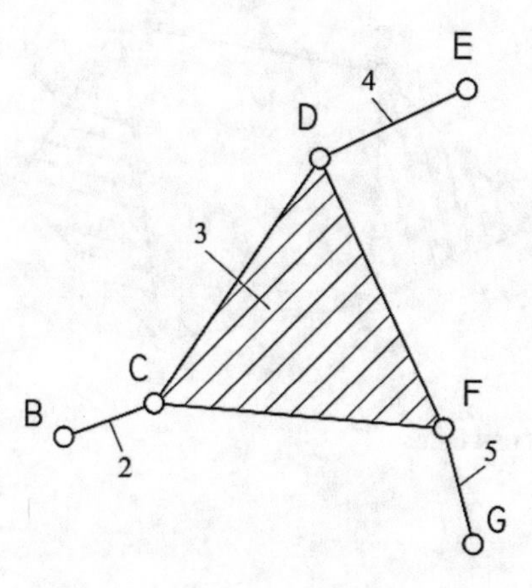

Fig. 1.25 Triad of the third order

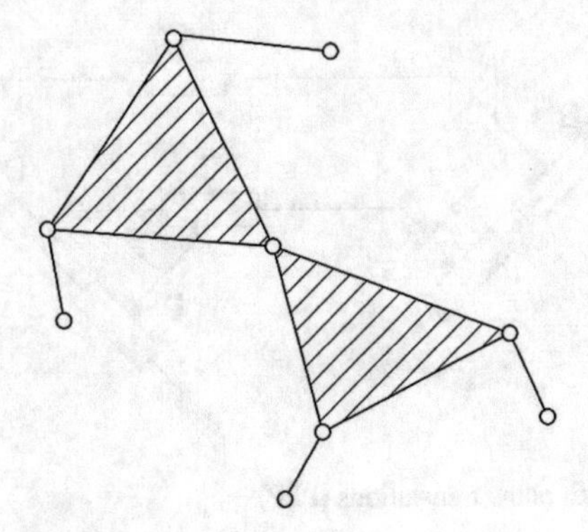

Fig. 1.26 Triad of the fourth order

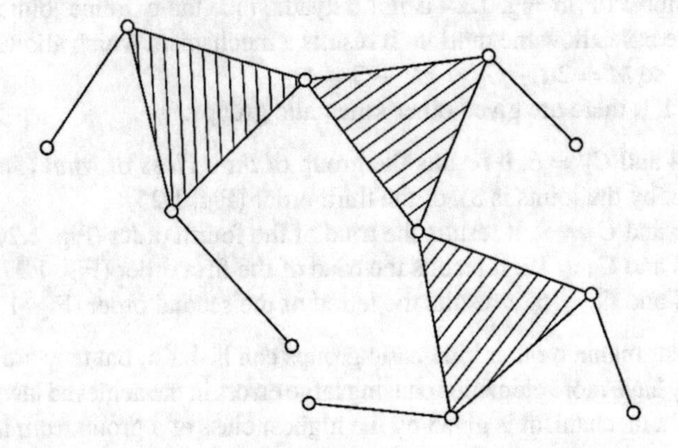

Fig. 1.27 Triad of the fifth order

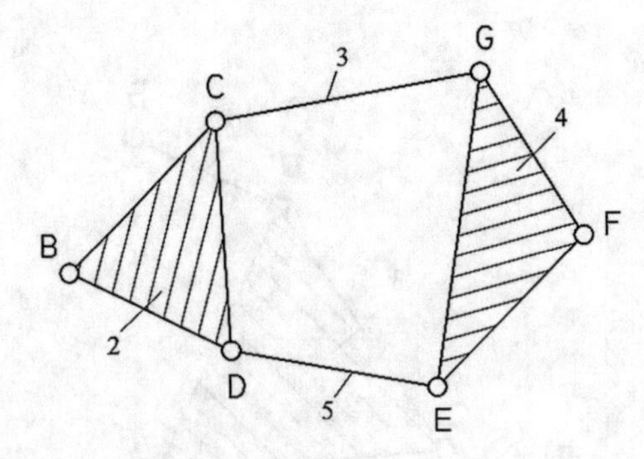

Fig. 1.28 Tetrad of the second order

1.5 Kinematic and Structural Diagrams

The kinematic diagrams represent the drawing of a mechanism to a certain scale. All the lengths are expressed at the same scale, but the angles have the real values. The kinematic elements are basically represented, without taking into consideration their width or depth [5].

The structural diagram shows how the elements are linked within the mechanism, in order to know the structure of the mechanism, namely the decomposition into its components. This diagram does not represent exactly either the lengths or the angles; the lengths that are equal to zero are being taken different from zero, and the sliders are replaced by revolute joints.

To illustrate these, it is considered a mechanism with the kinematic diagram in Fig. 1.29. Its structural diagram is given in Fig. 1.30.

If the driving link is considered to be the element 1 (Figs. 1.30 and 1.31a), then the structure of the mechanism includes a dyad PRR (Fig. 1.31b, c) and a dyad RRP (Fig. 1.31d, e).

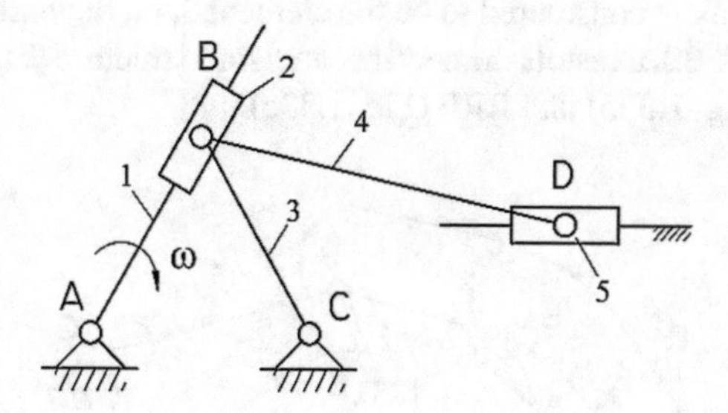

Fig. 1.29 Kinematic diagram of the mechanism

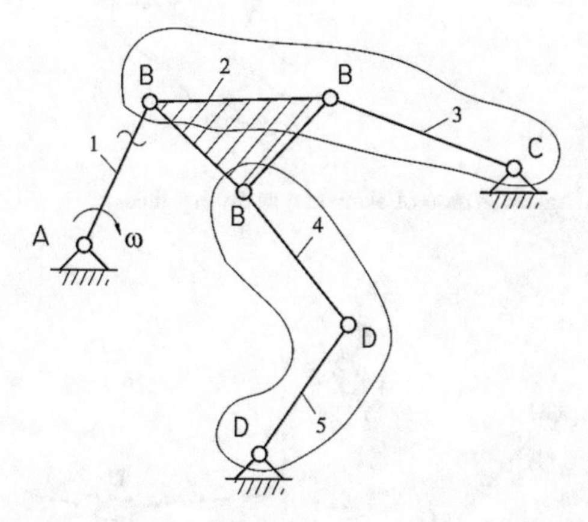

Fig. 1.30 Structural diagram of the mechanism

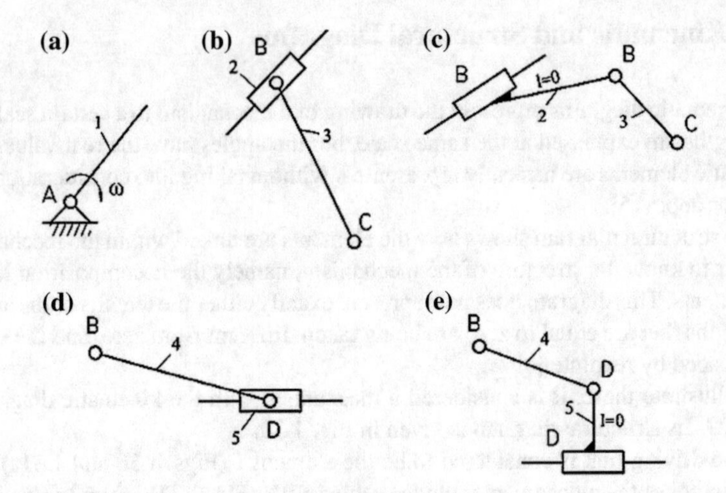

Fig. 1.31 Mechanism's structure in case of element 1 as driving link

If the driving link is considered to be the element 3, on the basis of the structural diagram from Fig. 1.32, it results the new mechanism's structure (Fig. 1.33), including the dyads RPR (Fig. 1.33b) and RRP (Fig. 1.33c).

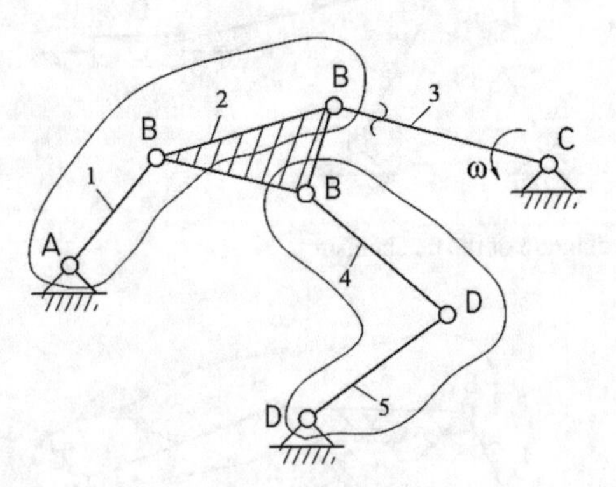

Fig. 1.32 Structural diagram in case of element 3 as driving link

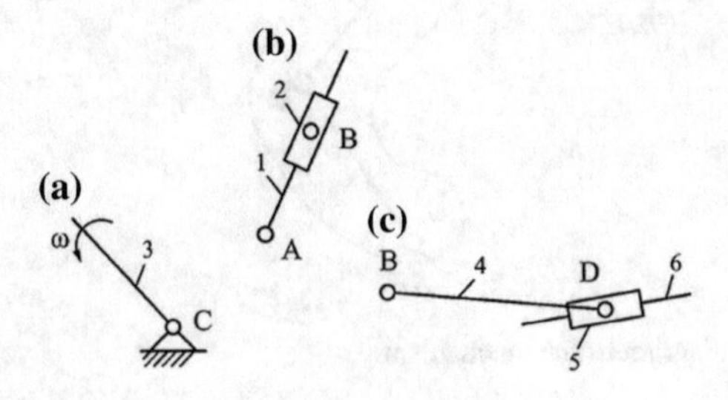

Fig. 1.33 Mechanism's structure in case of element 3 as driving link

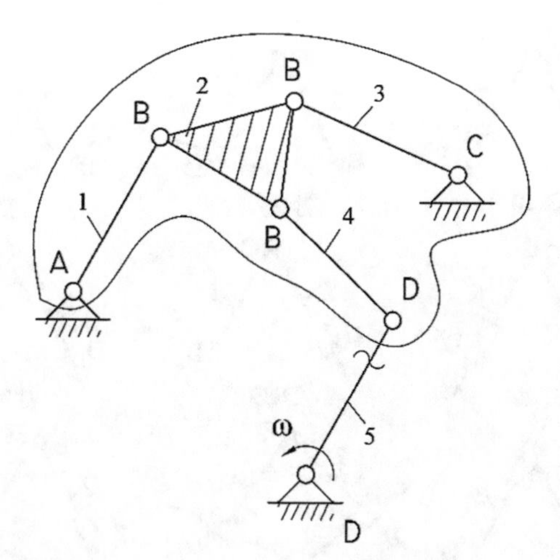

Fig. 1.34 Structural diagram in case of element 5 as driving link

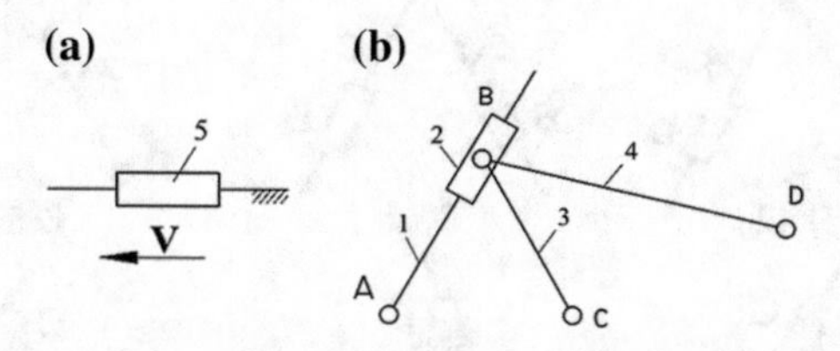

Fig. 1.35 Mechanism's structure in case of element 5 as driving link

It can be noticed that the mechanism's structure changed when the driving link has been changed.

If the kinematic element 5 is considered as the driving link, based on the structural diagram from Fig. 1.34, it results a different mechanism's structure, namely the driving link with translational motion (Fig. 1.35a) and a triad (Fig. 1.35b).

At the structural analysis, there were presented the dyads of five configurations. Below we will use the following symbols:

- *ET*, or *T*, or *P*—driving link with translational movement;
- *ER* or *R*—driving link with rotational movement;
- *MP*—driving link in plane movement;
- *D*10 or RRR—dyad of first configuration;
- *D*20 or RRP—dyad of second configuration;
- *D*30 or RPR—dyad of third configuration;
- *D*40 or PRP—dyad of fourth configuration;
- *D*50 or RPP—dyad of fifth configuration.

We made a more detailed classification of dyads [4], which takes into consideration the positions of prismatic pairs and is given in Fig. 1.36.

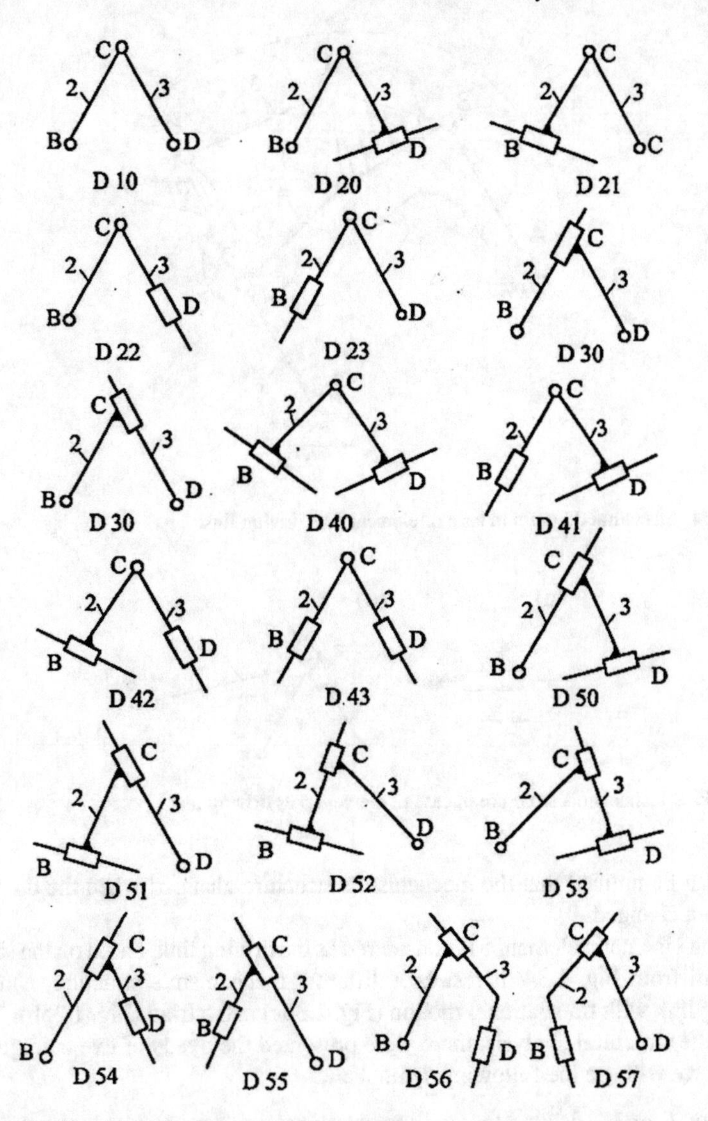

Fig. 1.36 Classification of dyads

References

1. Erdman AG, Sandor GN (1984) Mechanisms design. Prentice-Hall, Upper Saddle River, NJ
2. Myszka DH (1999) Machines and mechanisms. Prentice-Hall, Upper Saddle River, NJ
3. Popescu I (1997) Teoria mecanismelor şi a maşinilor. Sitech, Craiova
4. Popescu I, Ungureanu A (2000) Sinteza structurală şi cinematica a mecanismelor cu bare. Universitaria, Craiova
5. Popescu I, Călbureanu-Popescu MX (2017) Kinematics of planar mechanisms? nothing easier! Lambert Academic Publishing, Germany
6. Shigley JE, Uicker JJ (1995) Theory of machines and mechanisms. McGraw-Hill, New York

Chapter 2
Kinematic Analysis of Planar Mechanisms

Abstract Here, we describe the method of projection of the vector contours of the mechanisms on the reference system axes and the method of the distances from geometry. The discontinuities that occur in the kinematic analysis of the mechanisms are shown, represented by the jump from one trigonometric quadrant to another. The nonlinear algebraic system of equations from the calculation of the mechanism positions have two equal values, one positive and the other negative, resulted from the square roots. Further on, the kinematic calculations (positions, velocities, and accelerations) are given for the driving links, for the moving linkages with planar motion and for the RRR, RRP, RPR, PRP, and RPP dyads. These rules are exemplified in detail on the slider-crank mechanism, giving relationships, diagrams, and successive positions of the mechanism. Finally, the kinematic analysis of the common triad is done with the matrix method.

2.1 Closed-Loop Method

For the kinematic analysis of planar mechanisms, the vector loop equations are the most used [3, 11, 12]. The closed-loop method used in kinematics of the mechanisms is based on the projection of the vectorial loops on the axes of the system. In Fig. 2.1 are shown the $\overrightarrow{\boldsymbol{u}_i}$ vectors in the trigonometric quadrants.

General relationships can be written:

$$x_i = \boldsymbol{u}_i \cos \alpha_i \tag{2.1}$$

$$y_i = \boldsymbol{u}_i \sin \alpha_i \tag{2.2}$$

The angles are measured between the vectors and the positive direction of the x-axis. For mechanisms, the kinematic elements are considered as vectors and are projected on the x–y axes.

I. Popescu et al., *Mechanisms for Generating Mathematical Curves*,
Springer Tracts in Mechanical Engineering,
https://doi.org/10.1007/978-3-030-42168-7_2

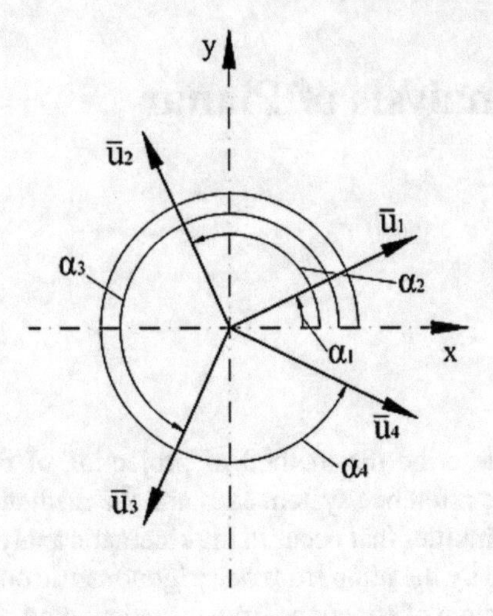

Fig. 2.1 Vectors in the four quadrants

Method of the distances

Another alternative method is the method of the distances [1], based on (2.3)–(2.5) [2]:

- The distance between two points:

$$(x - x_1)^2 + (y - y_1)^2 = d^2 \tag{2.3}$$

- The distance from a point to a line:

$$h = \frac{|Ax + By + C|}{\sqrt{a^2 + b^2}} \tag{2.4}$$

and

$$(x_2 - x_1)\sin\varphi - (y_2 - y_1)\cos\varphi = \pm h \tag{2.5}$$

where φ is the inclination angle of the line, x_1, y_1 are the coordinates of a point on the line, and x_2, y_2 are the coordinates of the starting point of the perpendicular to the line (Fig. 2.2).

Discontinuities

At the kinematic analysis of planar mechanisms, there are some situations that should be specified here. Therefore, on the diagrams representing the output characteristics (coordinates, angles, and displacement), sudden jumps may occur, as shown in Fig. 2.3.

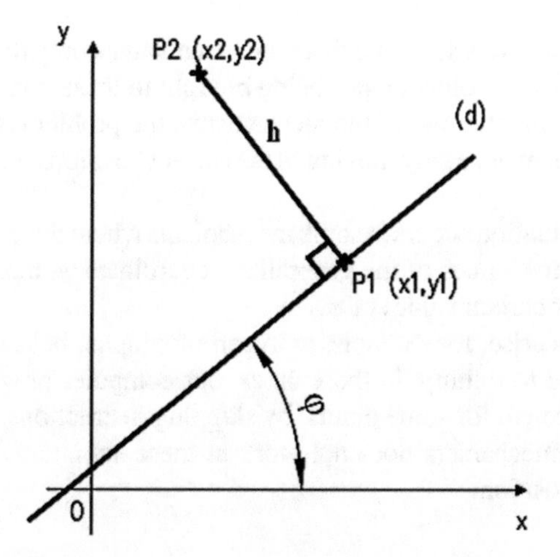

Fig. 2.2 Distance from a point to a line

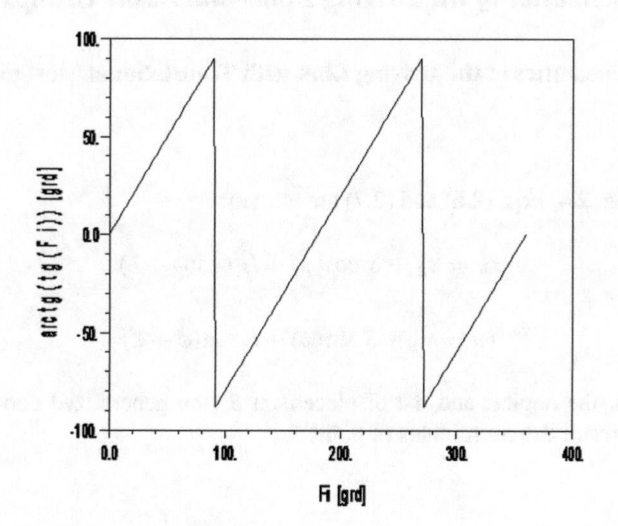

Fig. 2.3 Discontinuities in charts

A certain characteristic increases to a certain value, then decreases sharply in the negative area, increases again and decreases sharply, then increases again.

In the mechanism working cycle, such sudden jumps cannot occur. The explanation for this fact consists in the calculation of the angles based on the arctangent function, whose values are dependent on the quadrants.

In Fig. 2.3 on the abscissa is given the φ angle, which takes values between 0° and 360° and on the ordinate appear the values of the function arctan (tan φ), that should show all values of φ.

From the diagram, it is seen that there are given values only to φ, between 0° and 90° and 0 to (−90°), the other angles being brought to these quadrants.

In the effective functioning of the mechanisms, the problem is solved due to the inertia of the mechanism, easily moving to the correct positions that follow according to the cycle.
Another similar situation occurs at some mechanisms where the coordinates increase rapidly around some values of the generalized coordinate φ, meaning those curves tend to infinity for certain values of φ.

Such situations arise, for example, at hyperbolographe, because the branches of the hyperbola tend to infinity. In these cases, our computer programs have limited the coordinates growth for some points, by skipping instructions.

Currently, the mechanism does not work in these subintervals, so it should be fitted in another position.

2.1.1 *Kinematics of the Driving Links and Assur Groups*

2.1.1.1 Kinematics of the Driving Link with Translational Movement

Positions

Based on Fig. 2.4, Eqs. (2.6) and (2.7) are written:

$$x_B = x_A + S\ \cos(\varphi) + l_1\ \cos(\varphi + \lambda) \tag{2.6}$$

$$y_B = y_A + S\ \sin(\varphi) + l_1\ \sin(\varphi + \lambda) \tag{2.7}$$

The lengths, the angles, and the displacement S (the generalized coordinate) are known and result the coordinates of point B.

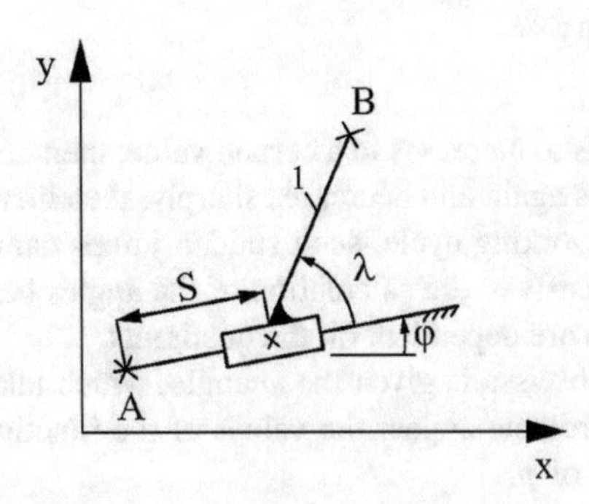

Fig. 2.4 Driving link with translational movement

Velocities

Taking the derivative with respect to time of the relations for positions (2.6) (2.7), the components of the velocity for point B are obtained:

$$\dot{x}_B = \dot{S}\cos(\varphi) \tag{2.8}$$

$$\dot{y}_B = \dot{S}\sin(\varphi) \tag{2.9}$$

It is known $\dot{S}$ as being the velocity of the slider in the translational motion.

Accelerations

Taking the derivative with respect to time of the relations for velocities (2.8).and (2.9), the components of the acceleration for point B are obtained:

$$\ddot{x}_B = \ddot{S}\cos(\varphi) \tag{2.10}$$

$$\ddot{y}_B = \ddot{S}\sin(\varphi) \tag{2.11}$$

It is known $\ddot{S}$ as being the acceleration of the slider in the translational motion.

2.1.1.2 Kinematics of the Driving Link with Rotational Movement

Positions

Based on Fig. 2.5, we write (2.12) and (2.13):

$$x_B = x_A + l_1\cos(\varphi) \tag{2.12}$$

$$y_B = y_A + l_1\sin(\varphi) \tag{2.13}$$

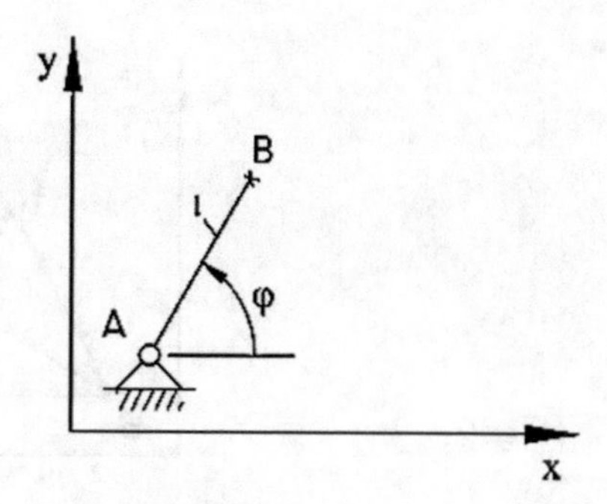

Fig. 2.5 Driving link with rotational movement

There are known the length l_1, x_A, y_A, and φ angle (the generalized coordinate) and the coordinates for the point B are obtained.

Velocities

Taking the derivative with respect to time of the relations for positions (2.12) and (2.13), the components of the velocity for point B are obtained:

$$\dot{x}_B = -l \sin(\varphi) \cdot \dot{\varphi} \tag{2.14}$$

$$\dot{y}_B = l_1 \cos(\varphi) \cdot \dot{\varphi} \tag{2.15}$$

It is known $\dot{\varphi}$ as being the given angular velocity in the rotational movement.

Accelerations

Taking the derivative with respect to time of the relations for velocities (2.14) and (2.15), the components of the acceleration for point B are obtained:

$$\ddot{x}_B = -l_1 \cos(\varphi) \cdot \dot{\varphi}^2 - l_1 \sin(\varphi) \cdot \ddot{\varphi} \tag{2.16}$$

$$\ddot{y}_B = l_1 \sin(\varphi) \cdot \dot{\varphi}^2 + l_1 \cos(\varphi) \cdot \ddot{\varphi} \tag{2.17}$$

It is known $\ddot{\varphi}$ as being the input angular acceleration.

2.1.1.3 Kinematics of the Linkage with Planar Movement

Positions. Displacements

Based on Fig. 2.6, we write Eqs. (2.18–2.21):

$$x_C = x_B + l_2 \cos \alpha \tag{2.18}$$

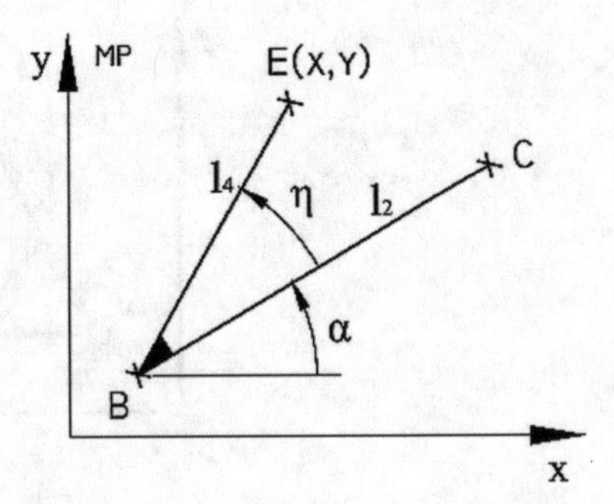

Fig. 2.6 Linkage with planar movement

$$y_C = y_B + l_2 \sin \alpha \tag{2.19}$$

$$x_E = x_B + l_4 \cos(\alpha + \eta) \tag{2.20}$$

$$y_E = y_B + l_4 \sin(\alpha + \eta) \tag{2.21}$$

There are known the lengths of the linkages: l_2, l_4, the coordinates of point B: x_B, y_B, the angles α and η and it results the coordinates for points C and E: x_C, y_C, x_E, y_E.

Velocities

Taking the derivative with respect to time of the relations for positions (2.18–2.21), it results:

$$\dot{x}_C = \dot{x}_B - l_2 \sin \alpha \cdot \dot{\alpha} \tag{2.22}$$

$$\dot{y}_C = \dot{y}_B + l_2 \cos \alpha \cdot \dot{\alpha} \tag{2.23}$$

$$\dot{x}_E = \dot{x}_B - l_4 \sin(\alpha + \eta) \cdot \dot{\alpha} \tag{2.24}$$

$$\dot{y}_E = \dot{y}_B + l_4 \cos(\alpha + \eta) \cdot \dot{\alpha} \tag{2.25}$$

There are known the components of the velocity for one point and the angular velocity, or the components of the velocities for two points, resulting the components of the velocity for any point.

Accelerations

Taking the derivative with respect to time of the relations for velocities (2.22–2.25), it results:

$$\ddot{x}_C = \ddot{x}_B - l_2 \cos \alpha \cdot \dot{\alpha}^2 - l_2 \sin \alpha \cdot \ddot{\alpha} \tag{2.26}$$

$$\ddot{y}_C = \ddot{y}_B - l_2 \sin \alpha \cdot \dot{\alpha}^2 + l_2 \cos \alpha \cdot \ddot{\alpha} \tag{2.27}$$

$$\ddot{x}_E = \ddot{x}_B - l_4 \cos(\alpha + \eta) \cdot \dot{\alpha}^2 - l_4 \sin(\alpha + \eta) \cdot \ddot{\alpha} \tag{2.28}$$

$$\ddot{y}_E = \ddot{y}_B - l_4 \sin(\alpha + \eta) \cdot \dot{\alpha}^2 + l_4 \cos(\alpha + \eta) \cdot \ddot{\alpha} \tag{2.29}$$

The components of the acceleration for one point and the angular acceleration, or the components of the accelerations for two points are known, resulting in the components of the acceleration for any point.

2.1.1.4 Kinematics of the Dyads

D10 (RRR)

Positions. Displacements

The given data are: $x_B, y_B, x_D, y_D, l_2, l_3$.
The unknowns are: α, β, x_C, y_C.
Based on Fig. 2.7, we write Eqs. (2.30) and (2.31) [4, 5, 9]:

$$x_C = x_B + l_2 \cos\alpha = x_D + l_3 \cos\beta \tag{2.30}$$

$$y_C = y_B + l_2 \sin\alpha = y_D + l_3 \sin\beta \tag{2.31}$$

The equations system (2.30–2.31) is nonlinear.

Velocities

Taking the derivative with respect to time of the relations for positions (2.30) and (2.31), it results:

$$\dot{x}_C = \dot{x}_B - l_2 \sin\alpha \cdot \dot{\alpha} = \dot{x}_D - l_3 \sin\beta \cdot \dot{\beta} \tag{2.32}$$

$$\dot{y}_C = \dot{y}_B + l_2 \cos\alpha \cdot \dot{\alpha} = \dot{y}_D + l_3 \cos\beta \cdot \dot{\beta} \tag{2.33}$$

Are known: the positions and $\dot{x}_B, \dot{y}_B, \dot{x}_D, \dot{y}_D$.
Are required: $\dot{\alpha}, \dot{\beta}$. The equations system (2.32–2.33) is linear.

Accelerations

Taking the derivative with respect to time of the relations for velocities (2.32) and (2.33), it results:

$$\ddot{x}_C = \ddot{x}_B - l_2 \cos\alpha \cdot \dot{\alpha}^2 - l_2 \sin\alpha \cdot \ddot{\alpha} = \ddot{x}_D - l_3 \cos\beta \cdot \dot{\beta}^2 - l_3 \sin\beta \cdot \ddot{\beta} \tag{2.34}$$

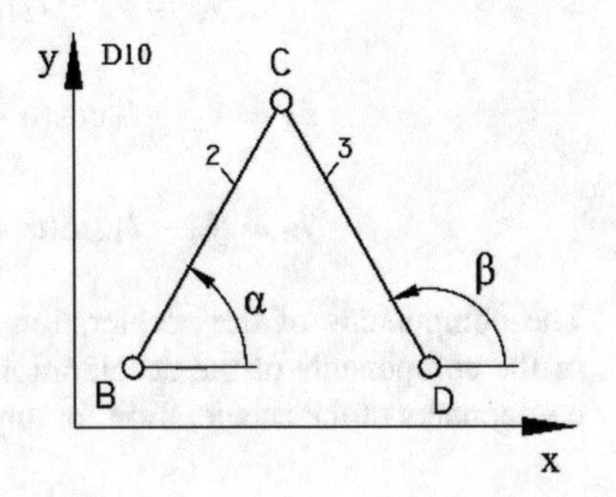

Fig. 2.7 RRR dyad (D10)

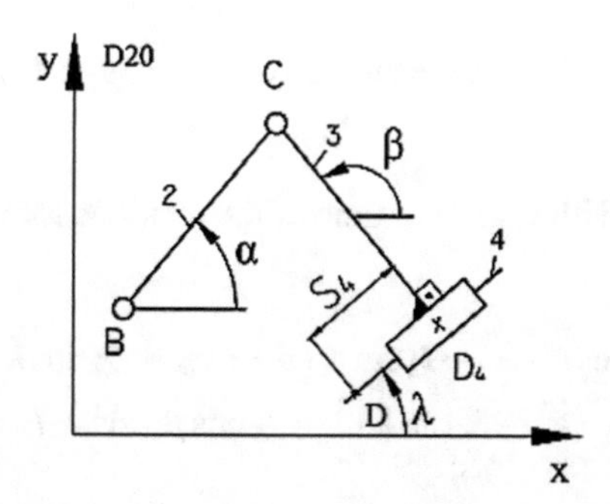

Fig. 2.8 RRP dyad (D20)

$$\ddot{y}_C = \ddot{y}_B - l_2 \sin\alpha \cdot \dot{\alpha}^2 + l_2 \cos\alpha \cdot \ddot{\alpha} = \ddot{y}_D - l_3 \sin\beta \cdot \dot{\beta}^2 + l_3 \cos\beta \cdot \ddot{\beta} \quad (2.35)$$

Are known: the positions, the velocities and $\ddot{x}_B, \ddot{y}_B, \ddot{x}_D, \ddot{y}_D$.
Are required: $\ddot{\alpha}$ and $\ddot{\beta}$. The equations system (2.34–2.35) is linear.

D20 (RRP)

Positions. Displacements

Based on Fig. 2.8, Eqs. (2.36–2.38) are written [4, 5, 9]:

$$x_C = x_B + l_2 \cos\alpha = x_D + S_4 \cos\lambda + l_3 \cos\beta \quad (2.36)$$

$$y_C = y_B + l_2 \sin\alpha = y_D + S_4 \sin\lambda + l_3 \sin\beta \quad (2.37)$$

$$\beta = \lambda + \pi/2 \quad (2.38)$$

Are known: $x_B, y_B, x_D, y_D, l_2, l_3, \lambda$.
Are required: α, S_4, x_C, y_C. The equations system (2.36–2.38) is nonlinear.

Velocities

Taking the derivative with respect to time of the relations for positions (2.36–2.38), it results:

$$\dot{x}_C = \dot{x}_B - l_2 \sin\alpha \cdot \dot{\alpha} = \dot{x}_D + \dot{S}_4 \cos\lambda - S_4 \sin\lambda \cdot \dot{\lambda} - l_3 \sin\beta \cdot \dot{\beta} \quad (2.39)$$

$$\dot{y}_C = \dot{y}_B + l_2 \cos\alpha \cdot \dot{\alpha} = \dot{y}_D + \dot{S}_4 \sin\lambda + S_4 \cos\lambda \cdot \dot{\lambda} + l_3 \cos\beta \cdot \dot{\beta} \quad (2.40)$$

$$\dot{\beta} = \dot{\lambda} \quad (2.41)$$

Are known: the positions and $\dot{x}_B, \dot{y}_B, \dot{x}_D, \dot{y}_D, \dot{\lambda}$.

Are required: $\dot{\alpha}$, $\dot{S}_4$, $\dot{x}_C$, $\dot{y}_C$.The equations system (2.39–2.41) is linear.

Accelerations

Taking the derivative with respect to time of the relations for velocities (2.39–2.41), it results:

$$\ddot{x}_C = \ddot{x}_B - l_2 \cos\alpha \cdot \dot{\alpha}^2 - l_2 \sin\alpha \cdot \ddot{\alpha} = \ddot{x}_D + \ddot{S}_4 \cos\lambda - 2\dot{S}_4 \sin\lambda \cdot \dot{\lambda}$$
$$- S_4 \cos\lambda \cdot \dot{\lambda}^2 - S_4 \sin\lambda \cdot \ddot{\lambda} - l_3 \cos\beta \cdot \dot{\beta}^2 - l_3 \sin\beta \cdot \ddot{\beta} \quad (2.42)$$

$$\ddot{y}_C = \ddot{y}_B - l_2 \sin\alpha \cdot \dot{\alpha}^2 + l_2 \cos\alpha \cdot \ddot{\alpha} = \ddot{y}_D + \ddot{S}_4 + 2\dot{S}_4 \cos\lambda \cdot \dot{\lambda}$$
$$- S_4 \sin\lambda \cdot \dot{\lambda}^2 + S_4 \cos\lambda \cdot \ddot{\lambda} - l_3 \sin\beta \cdot \dot{\beta}^2 + l_3 \cos\beta \cdot \ddot{\beta} \quad (2.43)$$

$$\ddot{\beta} = \ddot{\lambda} \quad (2.44)$$

Are known: the positions, the velocities, and $\ddot{x}_B$, $\ddot{y}_B$, $\ddot{x}_D$, $\ddot{y}_D$, $\ddot{\lambda}$.
Are required: $\ddot{\alpha}$, $\ddot{S}_4$, $\ddot{x}_C$, $\ddot{y}_C$. The equations system (2.42–2.44) is linear.

D30 (RPR)

Positions. Displacements

Based on Fig. 2.9, Eqs. (2.45–2.47) are written [4, 5, 9]:

$$x_C = x_B + S_2 \cos\alpha = x_D + l_3 \cos\beta \quad (2.45)$$

$$y_C = y_B + S_2 \sin\alpha = y_D + l_3 \sin\beta \quad (2.46)$$

$$\beta = \alpha + \pi/2 \quad (2.47)$$

Are known: x_B, y_B, x_D, y_D, l_3
Are required: α, β, S_2, x_C, y_C. The equations system (2.45–2.47) is nonlinear.

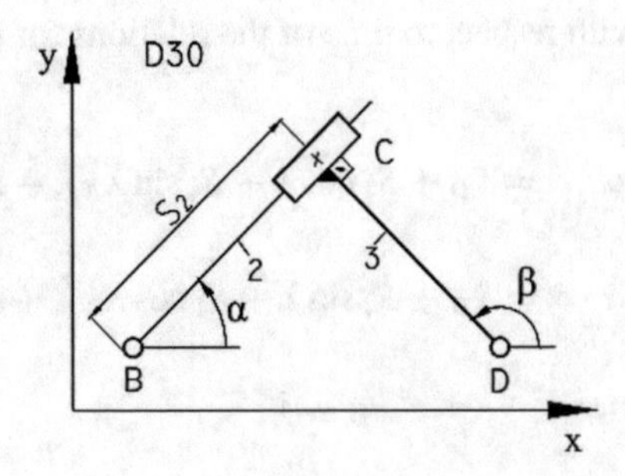

Fig. 2.9 RPR dyad (D30)

Velocities

Taking the derivative with respect to time of the relations for positions (2.45–2.47), it results:

$$\dot{x}_C = \dot{x}_B + \dot{S}_2 \cos\alpha - S_2 \sin\alpha \cdot \dot{\alpha} = \dot{x}_D - l_3 \sin\beta \cdot \dot{\beta} \quad (2.48)$$

$$\dot{y}_C = \dot{y}_B + \dot{S}_2 \sin\alpha + S_2 \cos\alpha \cdot \dot{\alpha} = \dot{y}_D + l_3 \cos\beta \cdot \dot{\beta} \quad (2.49)$$

$$\dot{\beta} = \dot{\alpha} \quad (2.50)$$

Are known: the positions and $\dot{x}_B, \dot{y}_B, \dot{x}_D, \dot{y}_D$
Are required: $\dot{\beta} = \dot{\alpha}, \dot{S}_2, \dot{x}_C, \dot{y}_C$. The equations system (2.48–2.50) is linear.

Accelerations

Taking the derivative with respect to time of the relations for velocities (2.48–2.50), it results:

$$\begin{aligned}\ddot{x}_C &= \ddot{x}_B + \ddot{S}_2 \cos\alpha - 2\dot{S}_2 \sin\alpha \cdot \dot{\alpha} - S_2 \cos\alpha \cdot \dot{\alpha}^2 - S_2 \sin\alpha \cdot \ddot{\alpha} \\ &= \ddot{x}_D - l_3 \cos\beta \cdot \dot{\beta}^2 - l_3 \sin\beta \cdot \ddot{\beta}\end{aligned} \quad (2.51)$$

$$\begin{aligned}\ddot{y}_C &= \ddot{y}_B + \ddot{S}_2 \sin\alpha + 2\dot{S}_2 \cos\alpha \cdot \dot{\alpha} - S_2 \sin\alpha \cdot \dot{\alpha}^2 + S_2 \cos\alpha \cdot \ddot{\alpha} \\ &= \ddot{y}_D - l_3 \sin\beta \cdot \dot{\beta}^2 - l_3 \cos\beta \cdot \ddot{\beta}\end{aligned} \quad (2.52)$$

$$\ddot{\beta} = \ddot{\alpha} \quad (2.53)$$

Are known: the positions, the velocities, and $\ddot{x}_B, \ddot{y}_B, \ddot{x}_D, \ddot{y}_D$.
Are required: $\ddot{\beta} = \ddot{\alpha}, \ddot{S}_2, \ddot{x}_C, \ddot{y}_C$. The equations system (2.51–2.53) is linear.

D40 (PRP)

Positions. Displacements

Based on Fig. 2.10, Eqs. (2.54) and (2.55) are written [4, 5, 9]:

$$x_C = x_B + S_1 \cos\gamma + l_2 \cos\alpha = x_D + S_4 \cos\lambda + l_3 \cos\beta \quad (2.54)$$

$$y_C = y_B + S_1 \sin\gamma + l_2 \sin\alpha = y_D + S_4 \sin\lambda + l_3 \sin\beta \quad (2.55)$$

Are known: $x_B, y_B, x_D, y_D, l_2, l_3, \gamma, \lambda$.
Are required: $\alpha, \beta, S_1, S_4, x_C, y_C$. The equations system (2.54–2.55) is linear.

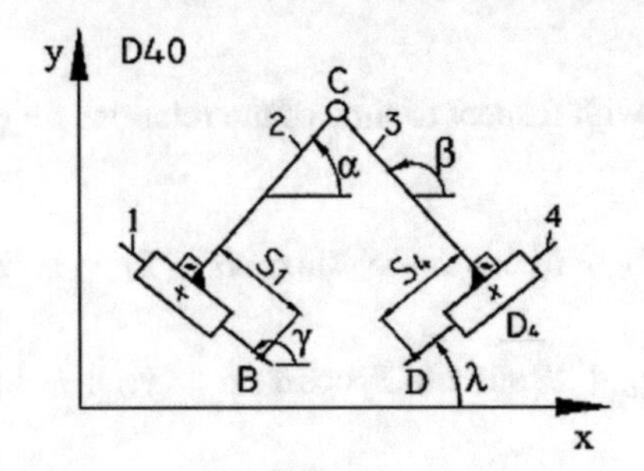

Fig. 2.10 PRP dyad (D40)

Velocities

Taking the derivative with respect to time of the relations for positions (2.54–2.55), it results:

$$\begin{aligned}\dot{x}_C &= \dot{x}_B + \dot{S}_1 \cos\gamma - S_1 \sin\gamma \cdot \dot{\gamma} - l_2 \sin\alpha \cdot \dot{\alpha} \\ &= \dot{x}_D + \dot{S}_4 \cos\lambda - S_4 \sin\lambda \cdot \dot{\lambda} - l_3 \sin\beta \cdot \dot{\beta}\end{aligned} \tag{2.56}$$

$$\begin{aligned}\dot{y}_C &= \dot{y}_B + \dot{S}_1 \sin\gamma + S_1 \cos\gamma \cdot \dot{\gamma} + l_2 \cos\alpha \cdot \dot{\alpha} \\ &= \dot{y}_D + \dot{S}_4 \sin\lambda + S_4 \sin\lambda \cdot \dot{\lambda} + l_3 \cos\beta \cdot \dot{\beta}\end{aligned} \tag{2.57}$$

$$\dot{\alpha} = \dot{\gamma};\ \dot{\beta} = \dot{\lambda} \tag{2.58}$$

Are known: the positions and $\dot{x}_B, \dot{y}_B, \dot{x}_D, \dot{y}_D, \dot{\gamma}, \dot{\lambda}$.
Are required: $\dot{\alpha}, \dot{\beta}, \dot{S}_1, \dot{S}_4 \dot{x}_C, \dot{y}_C$. The equations system (2.56–2.58) is linear.

Accelerations

Taking the derivative with respect to time of the relations for velocities (2.56–2.58), it results:

$$\begin{aligned}\ddot{x}_C &= \ddot{x}_B + \ddot{S}_1 \cos\gamma - 2\dot{S}_1 \sin\gamma \cdot \dot{\gamma} - S_1 \cos\gamma \cdot \dot{\gamma}^2 - S_1 \sin\gamma \cdot \ddot{\gamma} - l_2 \cos\alpha \cdot \dot{\alpha}^2 \\ &\quad - l_2 \sin\alpha \cdot \ddot{\alpha} = \ddot{x}_D + \ddot{S}_4 \cos\lambda - 2\dot{S}_4 \sin\lambda \cdot \dot{\lambda} - S_4 \sin\lambda \cdot \dot{\lambda}^2 \\ &\quad - S_4 \sin\lambda \cdot \ddot{\lambda} - l_3 \cos\beta \cdot \dot{\beta}^2 - l_3 \sin\beta \cdot \ddot{\beta}\end{aligned} \tag{2.59}$$

$$\begin{aligned}\ddot{y}_C &= \ddot{y}_B + \ddot{S}_1 \sin\gamma + 2\dot{S}_1 \cos\gamma \cdot \dot{\gamma} - S_1 \sin\gamma \cdot \dot{\gamma}^2 + S_1 \cos\gamma \cdot \ddot{\gamma} \\ &\quad - l_2 \sin\alpha \cdot \dot{\alpha}^2 + l_2 \cos\alpha \cdot \ddot{\alpha} = \ddot{y}_D + \ddot{S}_4 \sin\lambda + 2\dot{S}_4 \cos\lambda \cdot \dot{\lambda} \\ &\quad - S_4 \sin\lambda \cdot \dot{\lambda}^2 + S_4 \cos\lambda \cdot \ddot{\lambda} - l_3 \sin\beta \cdot \dot{\beta}^2 - l_3 \cos\beta \cdot \ddot{\beta}\end{aligned} \tag{2.60}$$

$$\ddot{\alpha} = \ddot{\gamma};\ \ddot{\beta} = \ddot{\lambda} \tag{2.61}$$

Are known: the positions, the velocities, and $\ddot{x}_B, \ddot{y}_B, \ddot{x}_D, \ddot{y}_D, \ddot{\gamma}, \ddot{\lambda}$.

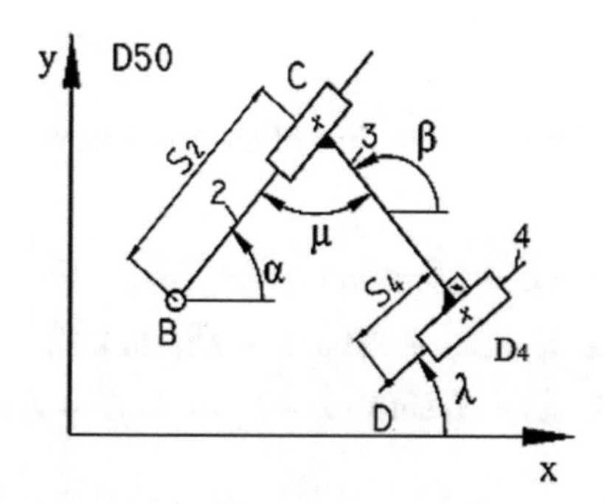

Fig. 2.11 RPP dyad (D50)

Are required: $\ddot{x}_C, \ddot{y}_C, \ddot{\alpha}, \ddot{\beta}, \ddot{S}_1, \ddot{S}_4$.The equations system (2.59–2.61) is linear.

D50(RPP)

Positions. Displacements

Based on Fig. 2.11, Eqs. (2.62–2.64) are written [4, 5, 9]:

$$x_c = x_B + S_2 \cdot \cos\alpha = x_D + S_4 \cdot \cos\lambda + l_3 \cos\beta \tag{2.62}$$

$$y_c = y_B + S_2 \cdot \sin\alpha = y_D + S_4 \cdot \sin\lambda + l_3 \sin\beta \tag{2.63}$$

$$\beta = \lambda + \pi/2; \alpha = \beta - \mu \tag{2.64}$$

Are known: $x_B, y_B, x_D, y_D, l_3, \mu, \lambda$
Are required: $\alpha, \beta, S_2, S_4, x_C, y_C$. The equations system (2.62–2.64) is linear.

Velocities

Taking the derivative with respect to time of the relations for positions (2.62–2.64), it results:

$$\begin{aligned}\dot{x}_C &= \dot{x}_B + \dot{S}_2 \cos\alpha - S_2 \sin\alpha \cdot \dot{\alpha} \\ &= \dot{x}_D + \dot{S}_4 \cos\lambda - S_4 \sin\lambda \cdot \dot{\lambda} - l_3 \sin\beta \cdot \dot{\beta}\end{aligned} \tag{2.65}$$

$$\begin{aligned}\dot{y}_C &= \dot{y}_B + \dot{S}_2 \sin\alpha + S_2 \cos\alpha \cdot \dot{\alpha} \\ &= \dot{y}_D + \dot{S}_4 \sin\lambda + S_4 \cos\lambda \cdot \dot{\lambda} + l_3 \cos\beta \cdot \dot{\beta}\end{aligned} \tag{2.66}$$

$$\dot{\beta} = \dot{\lambda}; \dot{\alpha} = \dot{\beta} \tag{2.67}$$

Are known: the positions and $\dot{x}_B, \dot{y}_B, \dot{x}_D, \dot{y}_D, \dot{\lambda}$.
Are required: $\dot{S}_2, \dot{S}_4, \dot{x}_C, \dot{y}_C$.The equations system (2.65–2.67) is linear.

Accelerations

Taking the derivative with respect to time of the relations for velocities (2.65–2.67), it results:

$$\begin{aligned}\ddot{x}_C &= \ddot{x}_B + \ddot{S}_2 \cos\alpha - 2\dot{S}_2 \sin\alpha \cdot \dot{\alpha} - S_2 \cos\alpha \cdot \dot{\alpha}^2 \\ &\quad - S_2 \sin\alpha \cdot \ddot{\alpha} = \ddot{x}_D + \ddot{S}_4 \cos\lambda - 2\dot{S}_4 \sin\lambda \cdot \dot{\lambda} \\ &\quad - S_4 \sin\lambda \cdot \dot{\lambda}^2 - S_4 \sin\lambda \cdot \ddot{\lambda} - l_3 \cos\beta \cdot \dot{\beta}^2 - l_3 \sin\beta \cdot \ddot{\beta}\end{aligned} \tag{2.68}$$

$$\begin{aligned}\ddot{y}_C &= \ddot{y}_B + \ddot{S}_2 \sin\alpha - 2\dot{S}_2 \cos\alpha \cdot \dot{\alpha} - S_2 \sin\alpha \cdot \dot{\alpha}^2 \\ &\quad + S_2 \cos\alpha \cdot \ddot{\alpha} = \ddot{y}_D + \ddot{S}_4 \sin\lambda + 2\dot{S}_4 \cos\lambda \cdot \dot{\lambda} \\ &\quad - S_4 \sin\lambda \cdot \dot{\lambda}^2 + S_4 \cos\lambda \cdot \ddot{\lambda} - l_3 \sin\beta \cdot \dot{\beta}^2 + l_3 \cos\beta \cdot \ddot{\beta}\end{aligned} \tag{2.69}$$

$$\ddot{\beta} = \ddot{\lambda};\, \ddot{\alpha} = \ddot{\beta} \tag{2.70}$$

Are known: the positions, the velocities, and $\ddot{x}_B, \ddot{y}_B, \ddot{x}_D, \ddot{y}_D, \ddot{\lambda}$.
Are required: $\ddot{x}_C, \ddot{y}_C, \ddot{S}_2, \ddot{S}_4$. The equations system (2.68–2.70) is linear.

2.1.2 *Kinematics of the Slider-Crank Mechanism*

Slider-crank mechanism

The slider-crank mechanism in Fig. 2.12 may have as driving link the crank AB (car engine crankshaft at startup), or the slider *C* (piston, while driving) [10].

There are considered the vectorial loop ABC, having AC as closure vector and the open contour ABE [8, 9].
To calculate positions, Eqs. (2.71) and (2.72) are written:

$$AB \cos\varphi + BC \cos\alpha = S \tag{2.71}$$

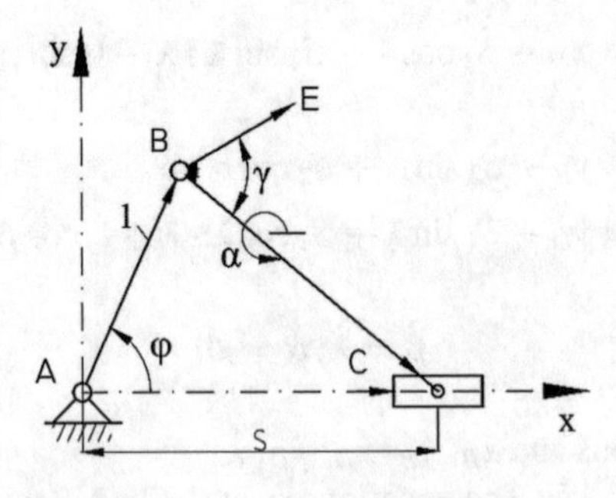

Fig. 2.12 Slider-crank mechanism

$$AB \sin \varphi + BC \sin \alpha = 0 \tag{2.72}$$

The coordinates of point E are calculated with Eqs. (2.73–2.74):

$$x_E = AB \cos \varphi + BE \cos (\alpha + \gamma) \tag{2.73}$$

$$y_E = AB \sin \varphi + BE \sin(\alpha + \gamma) \tag{2.74}$$

From Eqs. (2.71–2.74), the unknowns S, α, x_E, y_E are calculated.
A computer program has been developed based on (2.71–2.74) [7].

Figure 2.13 shows the mechanism in a certain position and the trajectory of the point *E*.
The initial input data have been adopted: *AB* = 50; *BC* = 100; *BE* = 40 (mm); and $\gamma = 45°$.

The successive positions of the mechanisms are presented in Fig. 2.14.

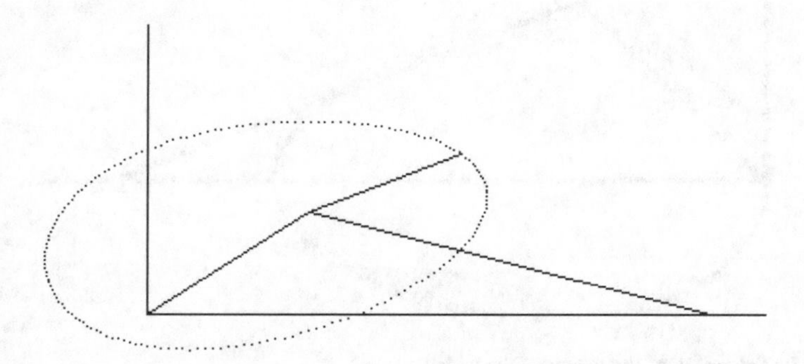

Fig. 2.13 Trajectory of the point *E*

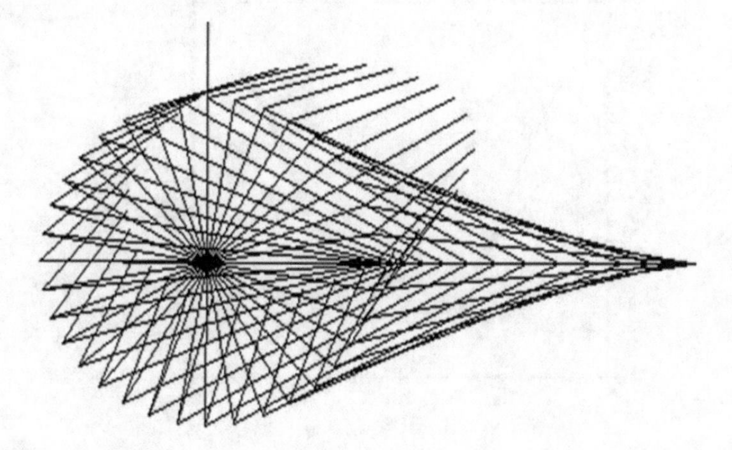

Fig. 2.14 Successive positions of the mechanism

If point E is the midpoint of the segment BC (Fig. 2.12), then the trajectory is obtained in Fig. 2.15.

In this way, the trajectories of any points belonging to the rod BC can be obtained.

The coordinates variations of point E, for the initial values, are given in Fig. 2.16. The curves have symmetries.

The actual values are given in Table 2.1.

The variation of the displacement S is shown in Fig. 2.17. This curve is also symmetrical.

Taking the derivative with respect to time of the relations for positions (2.71) and (2.72), there result the velocities:

$$-AB\sin\varphi\cdot\dot{\varphi} - BC\sin\alpha\cdot\dot{\alpha} = \dot{S} \tag{2.75}$$

$$AB\cos\varphi\cdot\dot{\varphi} + BC\cos\alpha\cdot\dot{\alpha} = 0 \tag{2.76}$$

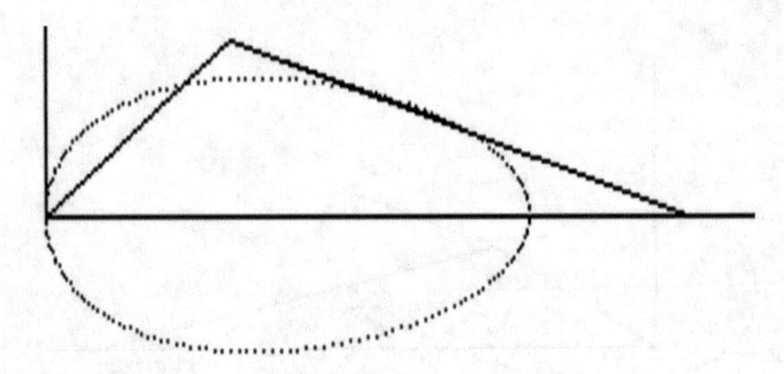

Fig. 2.15 Trajectory of the center of the slider

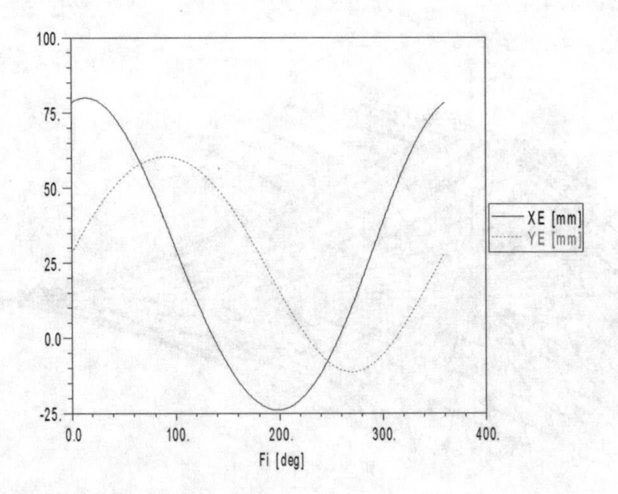

Fig. 2.16 Coordinates variations of point E

Table 2.1 Values for *XE* and *YE*

Fi (deg)	*XE* (mm)	*YE* (mm)
0	78.28429	28.28426
20	79.68916	40.13171
40	74.1763	49.83263
60	62.74259	56.54889
80	47.22744	59.93077
100	29.86264	59.93078
120	12.74263	56.54892
140	−2.428082	49.83269
160	−14.28005	40.13178
180	−21.71568	28.28434
200	−23.95389	15.60359
220	−20.60898	3.734753
240	−11.75243	−5.558666
260	2.007871	−10.6954
280	19.37264	−10.69545
300	38.24746	−5.558815
320	55.99534	3.734539
340	70.01523	15.60333
360	78.28423	28.28409

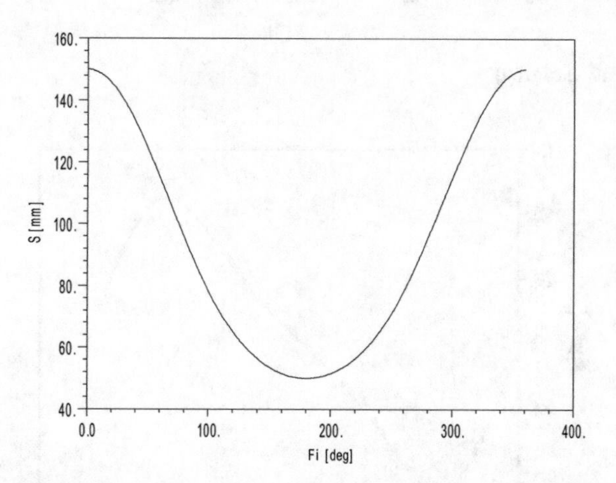

Fig. 2.17 Variation $S(\varphi)$

The resulting equations system (2.75–2.76) is linear, and the unknowns $\dot{\alpha}$ and $\dot{S}$ are calculated. The obtained curves are given in Figs. 2.18 and 2.19.
For E point (Fig. 2.12), it results:

$$\dot{x}_E = -AB \sin \varphi \cdot \dot{\varphi} - BE \sin(\alpha \cdot \gamma) \cdot \dot{\alpha} \tag{2.77}$$

$$\dot{y}_E = AB \cos \varphi \cdot \dot{\varphi} + BE \cos(\alpha + \gamma) \cdot \dot{\alpha} \tag{2.78}$$

In Fig. 2.20, the variations of velocity components of point E are given.

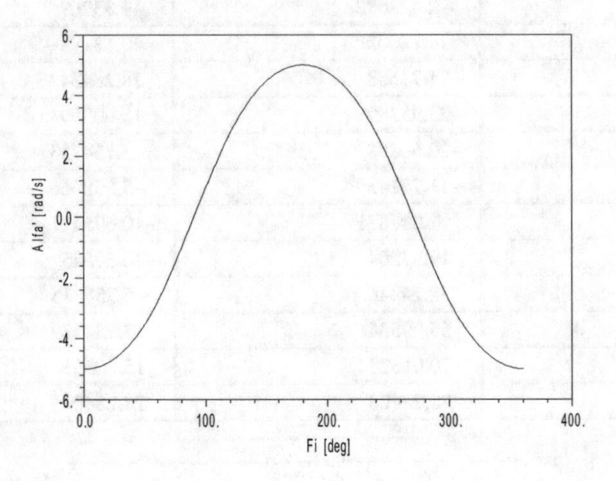

Fig. 2.18 Variation of $\dot{\alpha}(\varphi)$

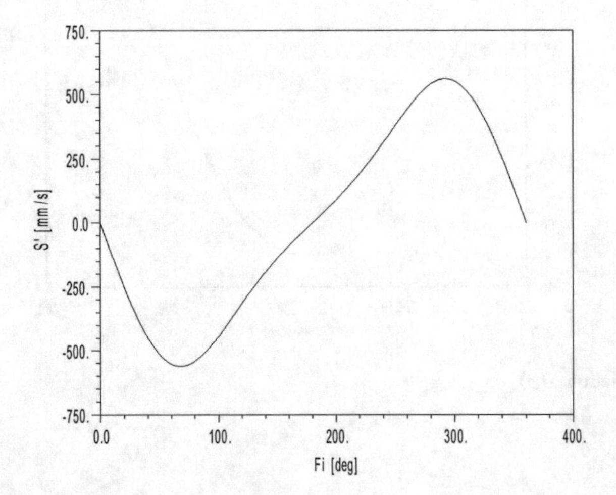

Fig. 2.19 Variation of $\dot{S}(\varphi)$

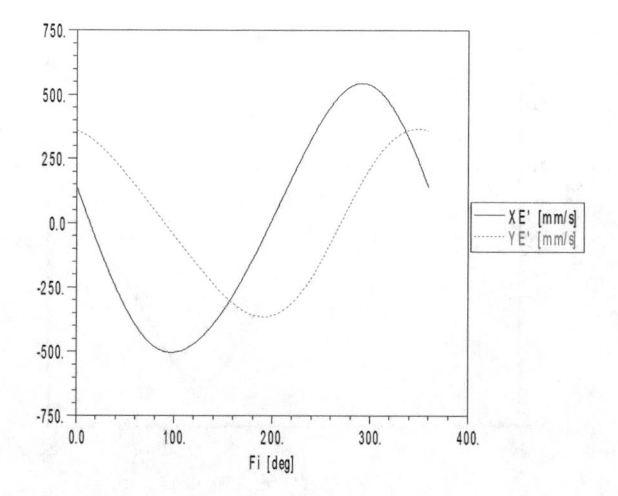

Fig. 2.20 Variations of velocity components for point E

To calculate the accelerations, the velocities relations (2.77) and (2.78) take the derivative with respect to time, resulting:

$$-AB\cos\varphi\cdot\dot{\varphi}^2 - AB\sin\varphi\cdot\ddot{\varphi} - BC\cos\alpha\cdot\dot{\alpha}^2 - BC\sin\alpha\cdot\ddot{\alpha} = \ddot{S} \tag{2.79}$$

$$-AB\sin\varphi\cdot\dot{\varphi}^2 + AB\cos\varphi\cdot\ddot{\varphi} - BC\sin\alpha\cdot\dot{\alpha}^2 + BC\cos\alpha\cdot\ddot{\alpha} = 0 \tag{2.80}$$

From Eqs. (2.79) and (2.80) are calculated the unknowns: $\ddot{\alpha}$ and $\ddot{S}$.
Similarly, for the accelerations of point E, are obtained Eqs. (2.81) and (2.82):

$$\ddot{x}_E = -AB\cos\varphi - \dot{\varphi}^2 - AB\sin\varphi - \ddot{\varphi} - BE\cos(\alpha+\gamma)\cdot\dot{\alpha}^2 - BE\sin(\alpha+\gamma)\cdot\ddot{\alpha} \tag{2.81}$$

$$\ddot{y}_E = -AB\sin\varphi\cdot\dot{\varphi}^2 + AB\cos\varphi\cdot\ddot{\varphi} - BE\sin(\alpha+\gamma)\cdot\dot{\alpha}^2 + BE\cos(\alpha+\gamma)\cdot\ddot{\alpha} \tag{2.82}$$

The results are given in Fig. 2.21 for $\ddot{\alpha}$ and in Fig. 2.22 for $\ddot{S}$.
Detailed results are given in Table 2.2.

2.2 Kinematics of Third Order Triad

Displacements

Equations (2.83–2.88) are written according to Fig. 2.23 [5]:

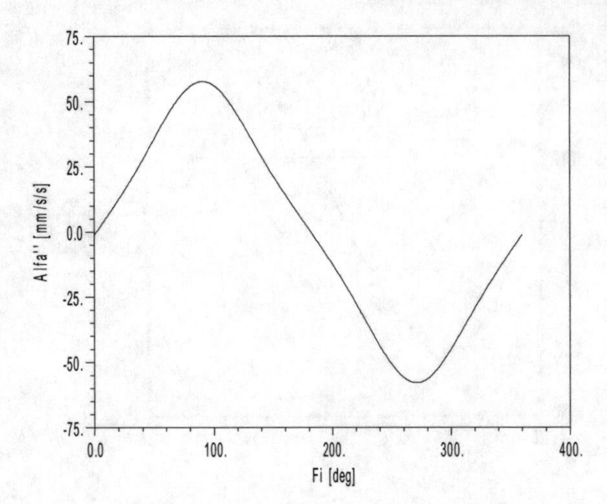

Fig. 2.21 Variation $\ddot{\alpha}(\varphi)$

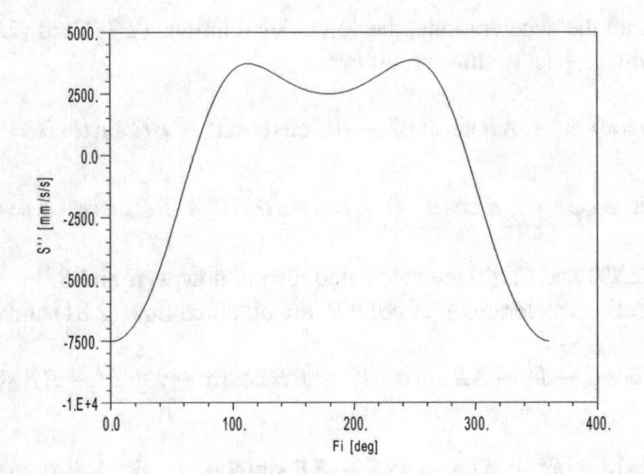

Fig. 2.22 Variation $\ddot{S}(\varphi)$

$$(x_4 - x_1)^2 + (y_4 - y_1)^2 = L_4^2 \tag{2.83}$$

$$(x_5 - x_2)^2 + (y_5 - y_2)^2 = L_2^2 \tag{2.84}$$

$$(x_6 - x_3)^2 + (y_6 - y_3)^2 = L_3^2 \tag{2.85}$$

$$(x_4 - x_5)^2 + (y_4 - y_5)^2 = L_5^2 \tag{2.86}$$

$$(x_4 - x_6)^2 + (y_4 - y_6)^2 = L_7^2 \tag{2.87}$$

Table 2.2 Results for slider motion (positions S, velocities $\dot{S}$, accelerations $\ddot{S}$)

Fi (deg)	S (mm)	$\dot{S}$ (mm/s)	$\ddot{S}$ (mm/s^2)
0	150	0	−7500
20	145.5116	−252.5597	−6760.219
40	132.9968	−451.3915	−4557.411
60	115.1388	−553.1088	−1383.889
80	95.71914	−541.5243	1694.854
100	78.35433	−443.2843	3450.988
120	65.13883	−312.9172	3664.159
140	56.39237	−191.3965	3155.041
160	51.54232	−89.46089	2669.332
180	50	−5.74E-04	2500
200	51.54227	89.45962	2705.11
220	56.39228	191.3949	3231.59
240	65.13869	312.9154	3789.321
260	78.35413	443.2826	3628.313
280	95.71889	541.5233	1911.498
300	115.1386	553.1093	−1162.6
320	132.9966	451.3935	−4376.817
340	145.5115	252.5629	−6659.172
360	150	3.44E-03	−7500

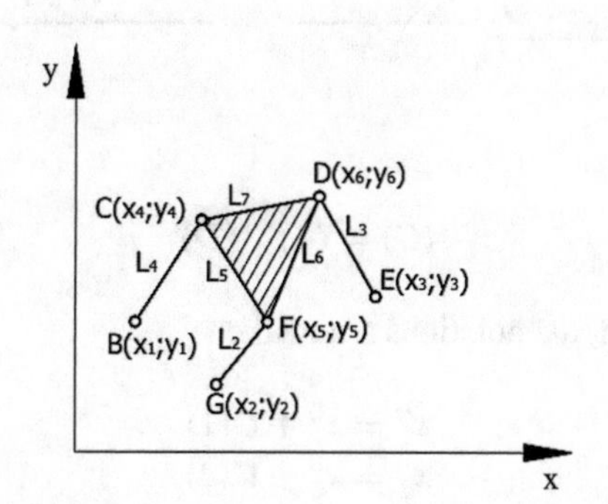

Fig. 2.23 Triad of the third order

$$(x_5 - x_6)^2 + (y_5 - y_6)^2 = L_6^2 \tag{2.88}$$

In these equations, the positions for the input points B, G, and E are known and the positions for the output points C, F, and D are the unknowns.

The nonlinear algebraic system is solved with the Newton–Raphson method. But it is necessary that the system's solutions could be initially approximated to around real solutions.
The algebraic system can be written under the form (2.89):

$$f_1(x_1, x_2 \ldots x_n) = a_i \quad i = 1, 2 \ldots n \tag{2.89}$$

Using the Taylor series, the algebraic system (2.89) is expanded:

$$\begin{aligned} F_i(x_1, x_2 \ldots x_n) &= f_i\left(x_1^0, x_2^0 \ldots x_n^0\right) \\ &\quad + \sum_{k=1}^{n} (x_k - x_k^0) \frac{\partial f_i}{\partial x_k} \left(x_1^0, x_2^0 \ldots x_n^0\right) + R \\ &= a_i \end{aligned} \tag{2.90}$$

where R is a function of higher-order derivatives terms and is insignificant if the initial solutions are close to the actual solutions.

Using the matrix form [6], it results:

$$\underbrace{\begin{pmatrix} \frac{\partial f_1}{\partial x_1^0} & \frac{\partial f_1}{\partial x_2^0} \cdots\cdots & \frac{\partial f_1}{\partial x_n^0} \\ \frac{\partial f_2}{\partial x_1^0} & \frac{\partial f_2}{\partial x_2^0} \cdots\cdots & \frac{\partial f_2}{\partial x_n^0} \\ \vdots & \vdots & \vdots \\ \frac{\partial f_n}{\partial x_1^0} & \frac{\partial f_n}{\partial x_2^0} \cdots\cdots & \frac{\partial f_n}{\partial x_n^0} \end{pmatrix}}_{(J)} \underbrace{\begin{pmatrix} x_1 - x_1^0 \\ x_1 - x_2^0 \\ \vdots \\ x_n - x_n^0 \end{pmatrix}}_{(C)} = \underbrace{\begin{pmatrix} a_1 - f_1(x_1^0, x_2^0 \cdots x_n^0) \\ a_2 - f_2(x_1^0, x_2^0 \cdots x_n^0) \\ \vdots \\ a_n - f_n(x_1^0, x_2^0, \cdots x_n^0) \end{pmatrix}}_{(D)} \tag{2.91}$$

or:

$$(C) = (J)^{-1}(D) \tag{2.92}$$

After the first iteration, the notations are made:

$$\begin{aligned} x_1^0 &= x_1^0 + C(1) \\ x_2^0 &= x_2^0 + C(2) \\ &\vdots \\ x_n^0 &= x_n^0 + C(n) \end{aligned} \tag{2.93}$$

so that, the difference $x_i - x_i^0$ is added to the initial value x_i^0, calculated at the previous iteration.

Using successive iterations, the calculation is continued until:

$$\left|x_i^0 - x_{i-1}^0\right| < \varepsilon \tag{2.94}$$

where ε is the allowable error.
For a specific case, it results:

$$\frac{\partial f_1}{\partial x_4^0} = J(1,1) = 2(x_4 - x_1); \quad \frac{\partial f_1}{\partial y_4^0} = J(1,2) = 2(y_4 - y_1) \tag{2.95}$$

respectively:

$$a_1 - f_1\left(x_1^0, x_2^0 \ldots x_n^0\right) = D(1,1) = L_4^2 - (x_4 - x_1)^2 - (y_4 - y_1)^2 \tag{2.96}$$

In a similar way, the other array elements are calculated and detailed in our computer program for solving the nonlinear algebraic system [7].

Velocities

Taking the derivative with respect to time of the initial system (2.83–2.88), it results:

$$(x_4 - x_1)(\dot{x}_4 - \dot{x}_1) + (y_4 - y_1)(\dot{y}_4 - \dot{y}_1) = 0 \tag{2.97}$$

$$(x_5 - x_2)(\dot{x}_5 - \dot{x}_2) + (y_5 - y_2)(\dot{y}_5 - \dot{y}_2) = 0 \tag{2.98}$$

$$(x_6 - x_3)(\dot{x}_6 - \dot{x}_3) + (y_6 - y_3)(\dot{y}_6 - \dot{y}_3) = 0 \tag{2.99}$$

$$(x_4 - x_5)(\dot{x}_4 - \dot{x}_5) + (y_4 - y_5)(\dot{y}_4 - \dot{y}_5) = 0 \tag{2.100}$$

$$(x_5 - x_6)(\dot{x}_5 - \dot{x}_6) + (y_5 - y_6)(\dot{y}_5 - \dot{y}_6) = 0 \tag{2.101}$$

$$(x_4 - x_6)(\dot{x}_4 - \dot{x}_6) + (y_4 - y_6)(\dot{y}_4 - \dot{y}_6) = 0 \tag{2.102}$$

In a matrix form, the equations system (2.97–2.102) is written as in (2.103):

$$\underbrace{\begin{pmatrix} x_4 - x_1 & y_4 - y_1 & 0 & 0 & 0 & 0 \\ 0 & 0 & x_5 - x_2 & y_5 - y_2 & 0 & 0 \\ 0 & 0 & 0 & 0 & x_6 - x_3 & y_6 - y_3 \\ x_4 - x_5 & y_4 - y_5 & x_5 - x_4 & y_5 - y_4 & 0 & 0 \\ 0 & 0 & x_5 - x_6 & y_5 - y_6 & x_6 - x_5 & y_6 - y_5 \\ x_4 - x_6 & y_4 - y_6 & 0 & 0 & x_6 - x_4 & y_6 - y_4 \end{pmatrix}}_{(N)} \underbrace{\begin{pmatrix} \dot{x}_4 \\ \dot{y}_4 \\ \dot{x}_5 \\ \dot{y}_5 \\ \dot{x}_6 \\ \dot{y}_6 \end{pmatrix}}_{(W)}$$

$$+\underbrace{\begin{pmatrix} x_1-x_4 & y_1-y_4 & 0 & 0 & 0 & 0 \\ 0 & 0 & x_2-x_5 & y_2-y_5 & 0 & 0 \\ 0 & 0 & 0 & 0 & x_3-x_6 & y_3-y_6 \\ 0 & 0 & 0 & 0 & 0 & 0 \\ 0 & 0 & 0 & 0 & 0 & 0 \\ 0 & 0 & 0 & 0 & 0 & 0 \end{pmatrix}}_{(M)} \underbrace{\begin{pmatrix} \dot{x}_1 \\ \dot{y}_1 \\ \dot{x}_2 \\ \dot{y}_2 \\ \dot{x}_3 \\ \dot{y}_3 \end{pmatrix}}_{(V)} = 0 \quad (2.103)$$

$$(W) = -(N)^{-1}(M)(V) \quad (2.104)$$

where (V) is the matrix of input velocities and (W) is the matrix of output velocities.

Accelerations

The system obtained for velocities (2.104) is derived with respect to time and results in (2.105):

$$(A2) = -(N)^{-1}[(M)(A) + (Q)] \quad (2.105)$$

where

$$(A2) = \begin{pmatrix} \ddot{x}_4 \\ \ddot{y}_4 \\ \ddot{x}_5 \\ \ddot{y}_5 \\ \ddot{x}_6 \\ \ddot{y}_6 \end{pmatrix} \quad (A) = \begin{pmatrix} \ddot{x}_1 \\ \ddot{y}_1 \\ \ddot{x}_2 \\ \ddot{y}_2 \\ \ddot{x}_3 \\ \ddot{y}_3 \end{pmatrix} \quad (Q) = \begin{pmatrix} (\dot{x}_4-\dot{x}_1)^2+(\dot{y}_4-\dot{y}_1)^2 \\ (\dot{x}_5-\dot{x}_2)^2+(\dot{y}_5-\dot{y}_2)^2 \\ (\dot{x}_6-\dot{x}_3)^2+(\dot{y}_6-\dot{y}_3)^2 \\ (\dot{x}_4-\dot{x}_5)^2+(\dot{y}_4-\dot{y}_5)^2 \\ (\dot{x}_5-\dot{x}_6)^2+(\dot{y}_5-\dot{y}_6)^2 \\ (\dot{x}_4-\dot{x}_6)^2+(\dot{y}_4-\dot{y}_6)^2 \end{pmatrix} \quad (2.106)$$

References

1. Cebâşev PL (1953) Izobrannâe trud. Izd. Nauka, Moskva
2. Creangă I (1951) ş.a. Curs de geometrie analitică. Editura Tehnică, Bucureşti
3. Pelecudi Chr (1975) Precizia mecanismelor. Editura Academiei, Bucureşti
4. Popescu I (1995) Mecanisme, vol. I. Universitatea din Craiova
5. Popescu I (1997) Teoria mecanismelor şi a maşinilor. Editura "SITECH", Craiova
6. Popescu I (1977) Mecanisme – analiza matriceala. Reprografia Univ. din Craiova
7. Popescu I (1997) Mecanisme. Noi algoritmi şi programe. Reprografia Universităţii din Craiova
8. Popescu I, Călbureanu – Popescu MX (2017) Kinematics of planar mechanisms? Nothing easier!. Lambert Academic Publishing, Germany
9. Popescu I, Ungureanu A (2000) Sinteza structurală şi cinematica a mecanismelor cu bare. Universitaria, Craiova
10. Reuleaux F (1963) The kinematics of machinery. Dover Publ, NY
11. Waldron KJ, Kinzel GL (1999) Kinematics, dynamics, and design of machinery. Wiley, New York
12. Wilson CE, Sadler JP (1991) Kinematics and dynamics of machinery. Harper Collins College Publishers, New York

Part II
Synthesis and Analysis of Mechanisms for Generating Mathematical Curves

Chapter 3
Mechanisms for Generating Straight Lines and Arcs

Abstract We analyze the Reuleaux mechanism that draws straight segments. There are relationships, diagrams, and positions of the mechanism which show that a point on a rod draws a straight line parallel to the ordinate of the axis system. A new mechanism whose point on the rod draws an arc is presented. There are also presented relationships, successive positions, and the resulting trajectory, including the arc. The velocities and accelerations are calculated, and their values are given in tables and diagrams.

3.1 An Approximate Rectilinear Guiding Mechanism

The linearization of mechanisms has been created starting with the seventeenth century, due to the occurrence of steam machines. Many mathematicians and engineers have contributed to this field: James Watt, Chebyshev, Paucellier, Reuleaux, and others [2, 7, 8].

Such a mechanism has to ensure an approximate rectilinear trajectory of a point belonging to the rod, without using a slider in the working area of the drawing point, a condition imposed from constructive reasons.

In Fig. 3.1, such a mechanism, created by Reuleaux [1, 6], is presented, where point C has to move vertically.

In order for point C to move vertically, it would have been easy to make the mechanism with the slider at C, type R-RRP. Due to constructive and functional reasons, the slider at C cannot be mounted. Consequently, the designer helped with an extra kinematic chain, thus making only an approximative synthesis.

3.1.1 The Synthesis of a New Mechanism

Given the mechanism in Fig. 3.1, it comes to build a rod-crank mechanism ABC, with a slider at C, so that it can draw the rectilinear trajectory of C. Next, using a graphical method, it starts with point C's positions and the slider at C is removed.

I. Popescu et al., *Mechanisms for Generating Mathematical Curves*,
Springer Tracts in Mechanical Engineering,
https://doi.org/10.1007/978-3-030-42168-7_3

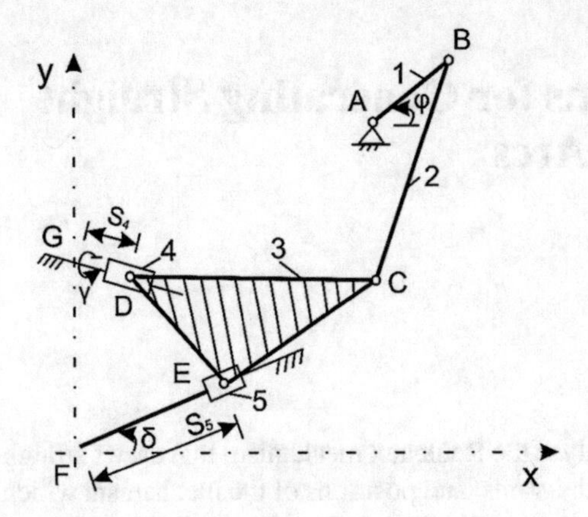

Fig. 3.1 Guiding mechanism

Another kinematic chain, based on rotational joints, aims to make sure that the wanted trajectory stays rectilinear. The construction is given in Fig. 3.2.

From the successive points of the drawing point C (Fig. 3.2), the arcs 3, 2-4, 1-5, 6-8, and 7 have been drawn, with a convenient length for CD. The position of a fixed point G and a radius GD have been chosen, further drawing an arc which intersects

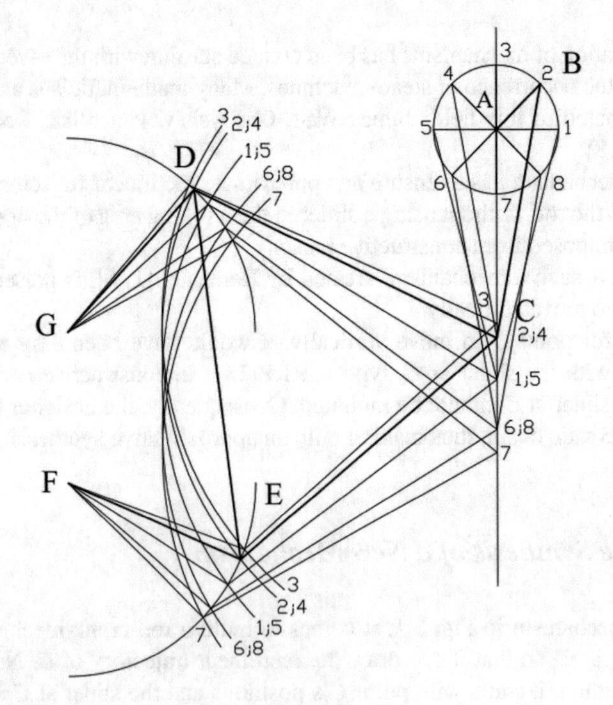

Fig. 3.2 Graphical construction of successive positions

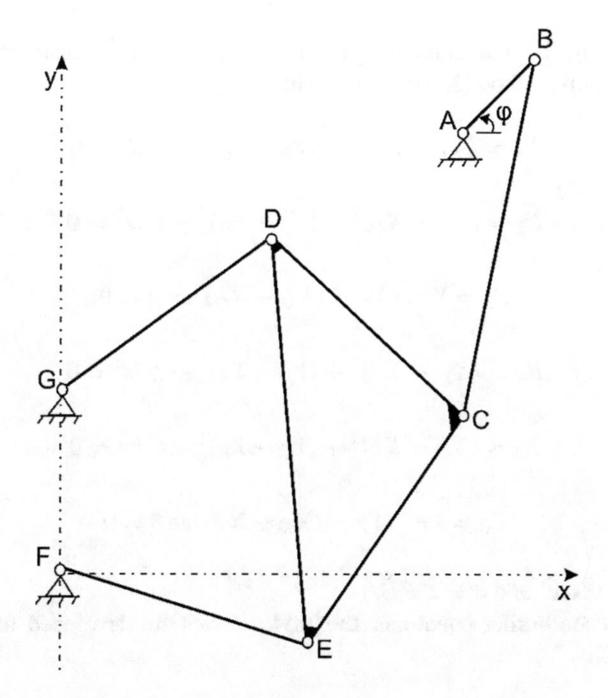

Fig. 3.3 New mechanism in a particular position

the precedent arcs at points D_i, $(i = \overline{1-8})$. Thus, it has been obtained the kinematic chain *CDG* with rotational joints at *C*, *D*, and *G*. The mechanism obtained so far, *ABCDG* has determined movement (meaning it is desmodromic), with mobility $M = 2$, a new restriction being necessary. For this purpose, point *F* is taken on the same vertical line with *G* and arc E_i, $(i = \overline{1-8})$ is drawn with radius $FE = GD$. Radius *CE* is equal to radius *CD*, because initial arcs, with *CD* radius passing in the lower part, are intersected with arcs E_i. A new kinematic chain *CEF* has been obtained. The obtained mechanism *ABCDGFE* has also the mobility $M = 2$, and thus, it is undesmodromic (does not have determined movement). It is found that the distance *DE* is constant so that, in the end, the desmodromic mechanism in Fig. 3.3 appears, with the ternary element *DCE*.

3.1.2 The Structure and Kinematics of the New Mechanism

The movements of input joints *B*, *G*, and *F* (external) are known, and the movements of output joints (internal) *C*, *D*, and *E* are required. The obtained equations system for positions is solved using a Newton–Raphson algorithm, after developing the Taylor series for each equation [4].

Based on Fig. 3.3, the relationships (3.1–3.6) (the method of distances combined with closed-loop method [3, 4]) are written:

$$E_1 = (X_B - X_C)^2 + (Y_B - Y_C)^2 - BC^2 = 0 \tag{3.1}$$

$$E_2 = (X_C - X_D)^2 + (Y_C - Y_D)^2 - CD^2 = 0 \tag{3.2}$$

$$E_3 = Y_D - Y_G - (X_D - X_G)\tan\gamma = 0 \tag{3.3}$$

$$E_4 = (X_D - X_E)^2 + (Y_D - Y_E)^2 - ED^2 = 0 \tag{3.4}$$

$$E_5 = (X_C - X_E)^2 + (Y_C - Y_E)^2 - CE^2 = 0 \tag{3.5}$$

$$E_6 = Y_E - Y_F - (X_E - X_F)\tan\delta = 0 \tag{3.6}$$

with $\gamma = \widehat{XGD}$ and $\delta = \widehat{XFE}$.

For the second-order equations, the Taylor series are developed for (3.1–3.6), based on (3.7):

$$E_i = E_i^0 + \frac{\partial E_i}{\partial X_C}\delta X_C + \frac{\partial E_i}{\partial Y_C}\delta Y_C + \frac{\partial E_i}{\partial X_D}\delta X_D + \frac{\partial E_i}{\partial Y_D}\delta Y_D + \frac{\partial E_i}{\partial X_E}\delta X_E + \frac{\partial E_i}{\partial Y_E}\delta Y_E \tag{3.7}$$

It results the matrix Eq. (3.8):

$$(P)_0 \begin{pmatrix} \delta X_C \\ \delta Y_C \\ \delta X_D \\ \delta Y_D \\ \delta X_E \\ \delta Y_E \end{pmatrix} = -\begin{pmatrix} E_1 \\ E_2 \\ E_3 \\ E_4 \\ E_5 \\ E_6 \end{pmatrix}_0 \tag{3.8}$$

Zero index refers to the initial position, which then becomes precedent, and matrix P has the form (3.9):

$$P = \begin{pmatrix} -2(X_B - X_C) & -2(Y_B - Y_C) & 0 & 0 & 0 & 0 \\ 2(X_C - X_D) & 2(Y_C - Y_D) & -2(X_C - X_D) & -2(Y_C - Y_D) & 0 & 0 \\ 0 & 0 & -\tan\gamma & 1 & 0 & 0 \\ 0 & 0 & 2(X_D - X_E) & 2(Y_D - Y_E) & -2(X_D - X_E) & -2(Y_D - Y_E) \\ 2(X_C - X_E) & 2(Y_C - Y_E) & 0 & 0 & -2(X_C - X_E) & -2(Y_C - Y_E) \\ 0 & 0 & 0 & 0 & \tan\delta & 1 \end{pmatrix} \tag{3.9}$$

After the first iteration, the new values of the coordinates of points C, D, and E are established (3.10–3.15):

$$X_C = X_C + \delta X_C \tag{3.10}$$

$$Y_C = Y_C + \delta Y_C \tag{3.11}$$

$$X_D = X_D + \delta X_D \tag{3.12}$$

$$Y_D = Y_D + \delta Y_D \tag{3.13}$$

$$X_E = X_E + \delta X_E \tag{3.14}$$

$$Y_E = Y_E + \delta Y_E \tag{3.15}$$

The calculus process continues until a wanted precision is reached.

3.1.3 Results Obtained

Measuring on Fig. 3.1 (on scale), the following dimensions of the mechanism (mm) have been resulted:

$Y_G = 38, X_A = 92, Y_A = 94, \varphi = 42, X_C = 91, Y_C = 31, X_D = 47, Y_D = 70, X_E = 56, Y_E = -17, GD = 59, CD = 58, AB = 22, BC = 80, CE = 58, FE = 58, ED = 88.$

In a particular position, the obtained mechanism is shown in Fig. 3.4.

The variations of the points C, D, and E coordinates are shown in Fig. 3.5, noting small variations of the point C's abscissa, which was the purpose.

In Fig. 3.6, the successive positions of the mechanism are shown, noting that this fulfills the imposed condition.

In Fig. 3.7, a detailed view of variations for the point C's abscissa and ordinate is given; the abscissa has an approximate linear variation, and the ordinate, a symmetrical curve, like in the rod-crank mechanism.

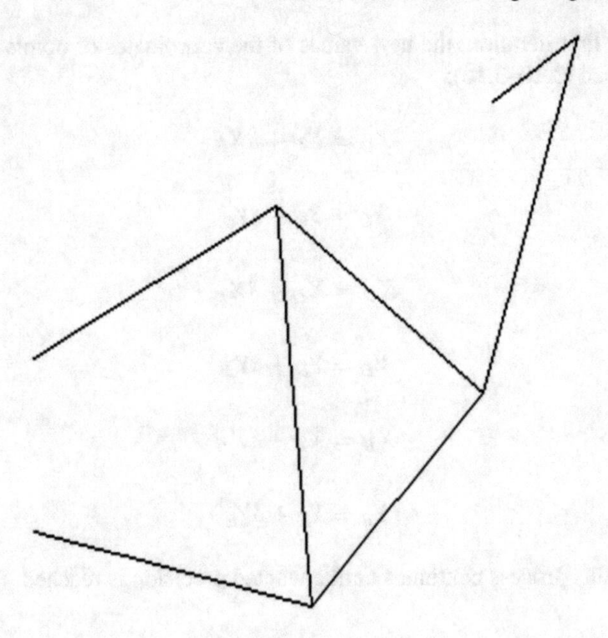

Fig. 3.4 The obtained mechanism in a particular position

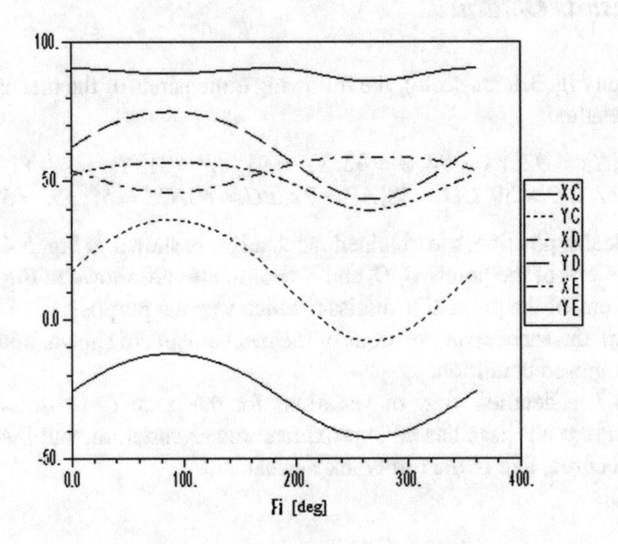

Fig. 3.5 Variations of the points C, D, and E coordinates

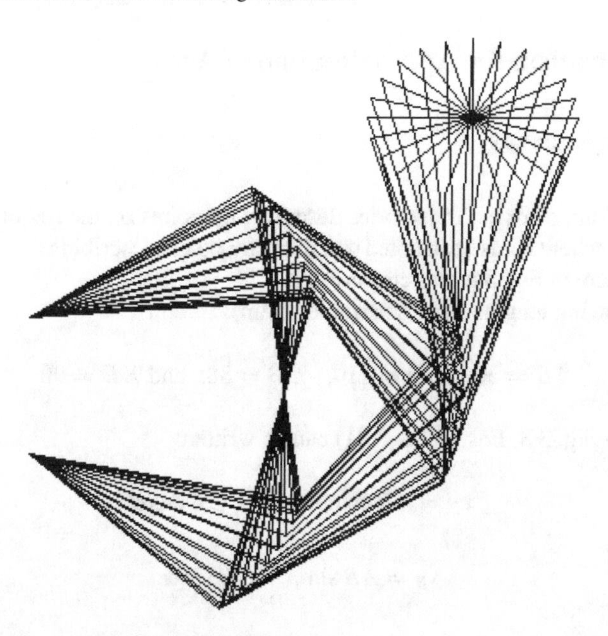

Fig. 3.6 Successive positions of the mechanism

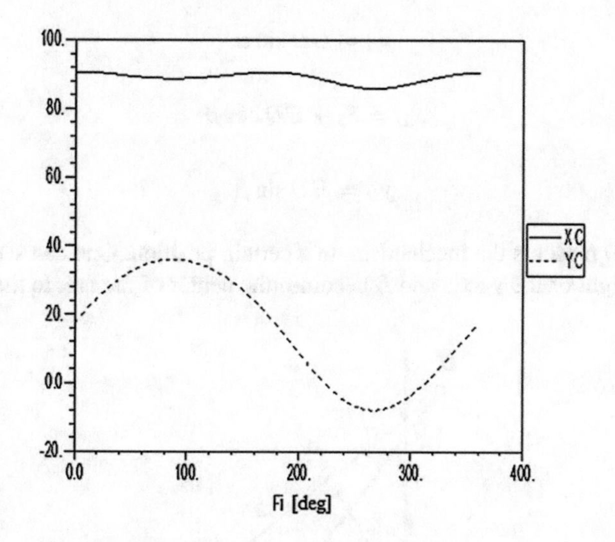

Fig. 3.7 Detalied view of variations for the point C coordinates

3.2 Mechanism for Approximation of Arcs

3.2.1 *Trajectories*

We started from studying trajectories described by points on the rod of a rod-crank mechanism, where it has been noted that there are points describing arcs. In this way, the mechanism in Fig. 3.8 was conceived [5].

The following lengths were determined (mm):

$$AB = 30;\ \ CD = 110;\ \ CB = 38;\ \text{and}\ ED = 90.$$

Based on Fig. 3.8, Eqs. (3.16–3.21) can be written:

$$x_B = AB\cos\varphi = S_3 + CB\cos\alpha \tag{3.16}$$

$$y_B = AB\sin\varphi + CB\sin\alpha \tag{3.17}$$

$$x_D = S_3 + CD\cos\alpha \tag{3.18}$$

$$y_D = CD\sin\alpha \tag{3.19}$$

$$x_D = S_5 + ED\cos\beta \tag{3.20}$$

$$y_D = ED\sin\beta \tag{3.21}$$

Figure 3.9 presents the mechanism, in a certain position. One can see that point C is to the right of the y-axis and E becomes the center of the arc, to the left of the

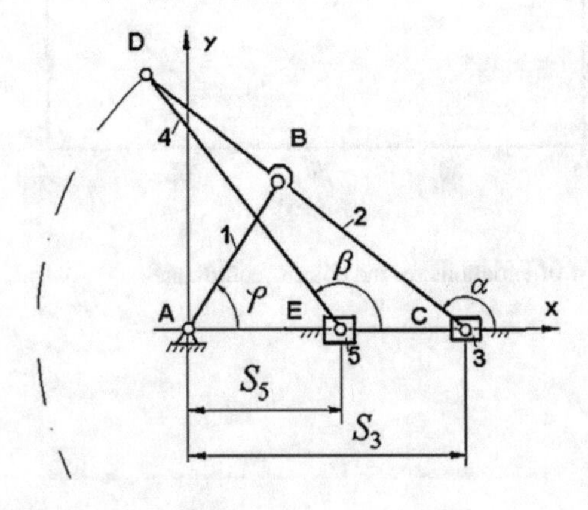

Fig. 3.8 Mechanism for generating an arc

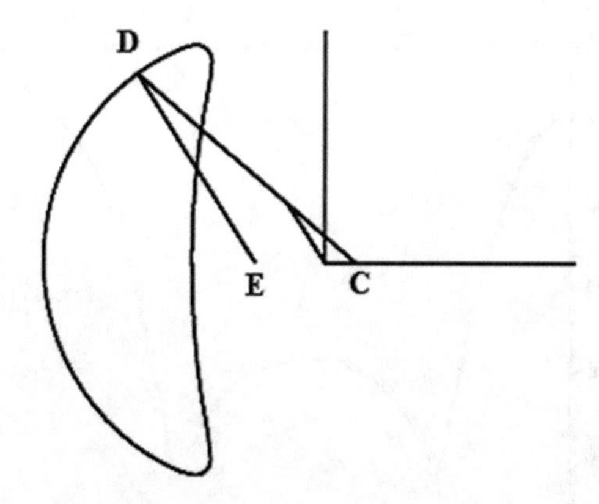

Fig. 3.9 The trajectory of the point D

y-axis. Another arc, drawn by point D in the movement of the mechanism on some subinterval, can also be noticed. In order to get a full rotation of the crank, the length of ED should be determined so as to prevent the blocking of the crank. Tests resulted in a value of $ED = 95$ mm. The arc drawn by point D presents major deviations with respect to the arc determined through the selection of three points on the specified area of the rod curve.

Figure 3.10 presents the successive positions of the rod ED. One can notice that E is a stationary point when point D draws the arc.

The variation of the displacement S_5 depending on the crank's angle is presented in Fig. 3.11. This diagram and Table 3.1 reveal that in the diagram's area corresponding to the interval $\varphi = 124° \ldots 135°$, S_5 presents small but still inacceptable variations.

Other values for ED were also tested. In Fig. 3.12, one can notice that the best value of S_5 is obtained for $ED = 110$ mm, and therefore, this value is considered in calculations for now on.

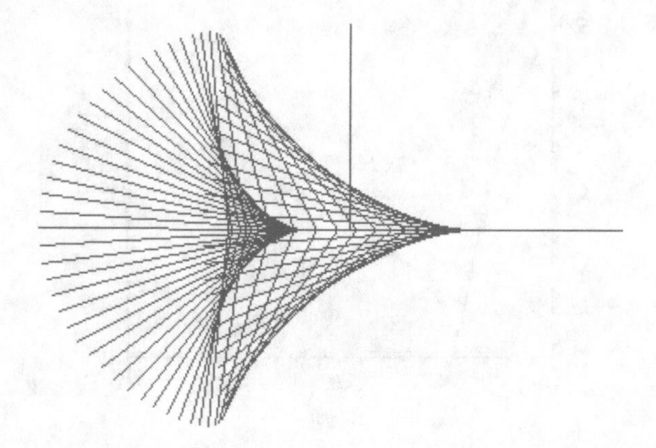

Fig. 3.10 Successive positions

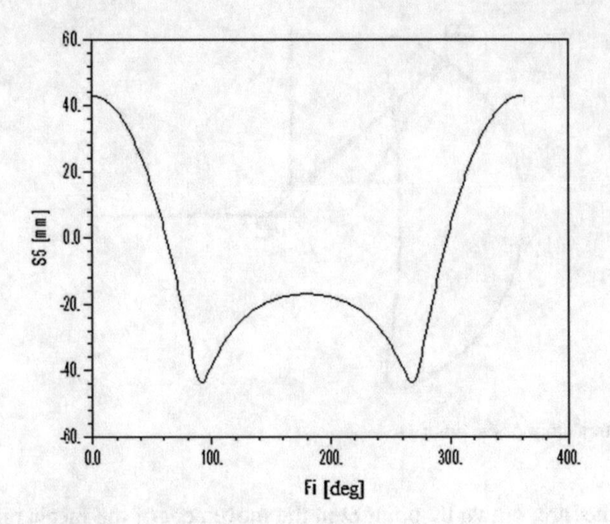

Fig. 3.11 Variation $S_5(\varphi)$

Table 3.1 Partial results for S_5

φ (°)	S_5 (mm)	φ (°)	S_5 (mm)
124	−25.32734	130	−23.28409
125	−24.95612	131	−22.98393
126	−24.59767	132	−22.69437
127	−24.25161	133	−22.4151
128	−23.91759	134	−22.14578
129	−23.59519	135	−21.88612

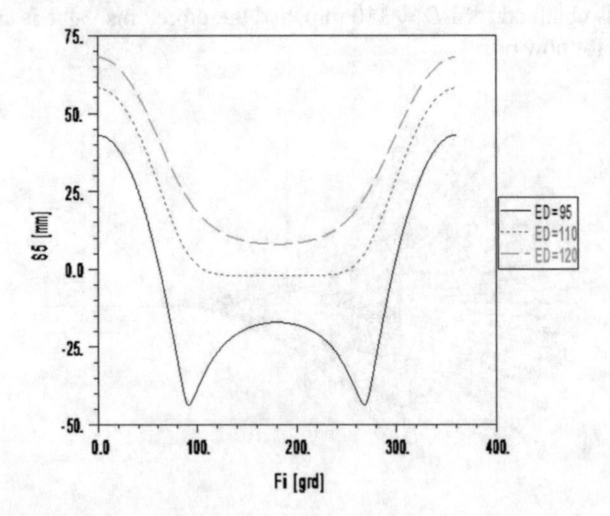

Fig. 3.12 Variation $S_5(\varphi)$ for different ED values

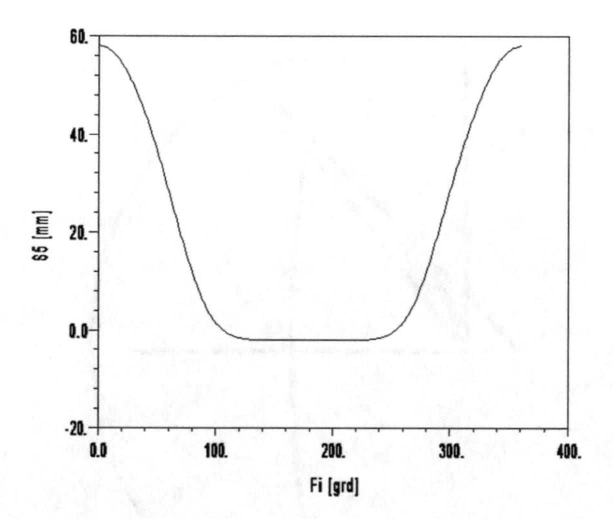

Fig. 3.13 Variation $S_5(\varphi)$ for $ED = 110$ mm

The curve $S_5(\varphi)$ for $ED = 110$ mm is presented in Fig. 3.13.

Almost constant values of the displacement S_5 can be seen (Fig. 3.13).

The initial position of the mechanism can also be placed in a different area, when, for the interval containing the arc, the point E is situated on the left side of the y-axis (Fig. 3.14).

Table 3.2 presents the detailed values of displacements S_5, velocities S_5', and accelerations S_5''.

For the case in Fig. 3.14, the curve $S_5(\varphi)$ obtained is presented in Fig. 3.15. It differs from that in Fig. 3.13, but it presents an almost constant horizontal area for $\varphi = 280° \ldots 360°$ and therefore it can conclude as beeing a useful alternative.

3.2.2 Velocities and Accelerations

To get the velocities (3.22–3.28), it is performed a time derivation of the equations corresponding to the positions of the mechanism (3.16–3.21):

$$AB = a; \quad BC = b; \quad CD = c; \quad ED = d \tag{3.22}$$

$$\dot{x}_B = -a \sin \varphi \cdot \dot{\varphi} = \dot{S}_3 - b \sin \alpha \cdot \dot{\alpha} \tag{3.23}$$

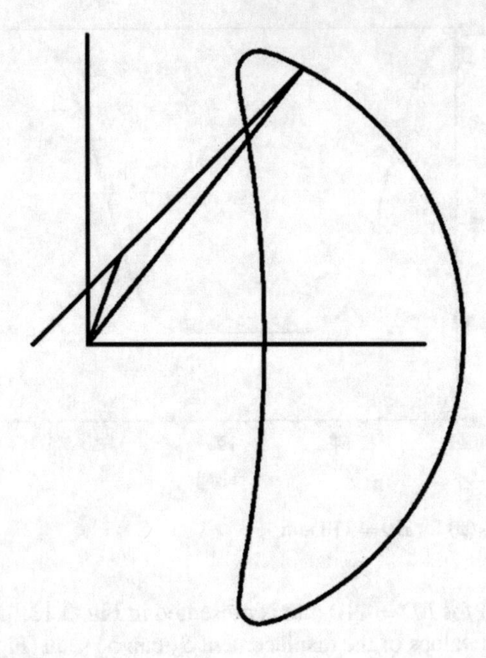

Fig. 3.14 Arc on the opposite side

$$\dot{y}_B = a\cos\varphi \cdot \dot{\varphi} = b\cos\alpha \cdot \dot{\alpha} \tag{3.24}$$

$$\dot{x}_D = \dot{S}_3 - c\sin\alpha \cdot \dot{\alpha} \tag{3.25}$$

$$\dot{y}_D = c\cos\alpha \cdot \dot{\alpha} \tag{3.26}$$

$$\dot{S}_5 - d\sin\beta \cdot \dot{\beta} = \dot{x}_D \tag{3.27}$$

$$d\cos\beta \cdot \dot{\beta} = \dot{y}_D \tag{3.28}$$

Taking the time derivative of velocities (3.22–3.28), the accelerations are obtained (3.29–3.34):

$$\ddot{x}_B = -a\cos\varphi \cdot \dot{\varphi}^2 - a\sin\varphi \cdot \ddot{\varphi} = \ddot{S}_3 - b\cos\alpha \cdot \dot{\alpha}^2 - b\sin\alpha \cdot \ddot{\alpha} \tag{3.29}$$

$$\ddot{y}_B = -a\sin\varphi \cdot \dot{\varphi}^2 + a\cos\varphi \cdot \ddot{\varphi} = -b\sin\alpha \cdot \dot{\alpha}^2 + b\cos\alpha \cdot \ddot{\alpha} \tag{3.30}$$

$$\ddot{x}_D = \ddot{S}_3 - c\cos\alpha \cdot \dot{\alpha}^2 - c\sin\alpha \cdot \ddot{\alpha} \tag{3.31}$$

$$\ddot{y}_D = -c\sin\alpha \cdot \dot{\alpha}^2 - c\cos\alpha \cdot \ddot{\alpha} \tag{3.32}$$

Table 3.2 Results for the slider 5's (Fig. 3.8) movement

φ (°)	S_5 (mm)	S_5' (mm/s)	S_5'' (mm/s^2)
120	−1.568848	−20.52344	718.5823
125	−1.793198	−12.35059	473.007
130	−1.925072	−7.028656	304.2777
135	−1.997368	−3.642609	190.9559
140	−2.032204	−1.548309	115.809
145	−2.044289	−0.3075562	66.3129
150	−2.043327	0.3720703	33.83756
155	−2.035645	0.6841431	12.60967
160	−2.025421	0.7566376	−1.166962
165	−2.015343	0.6747131	−9.950672
170	−2.007141	0.4960175	−15.30653
175	−2.001823	0.260952	−18.18251
180	−2	6.97E−06	−19.08814
185	−2.001823	−0.2609367	−18.18248
195	−2.01535	−0.6747131	−9.951348
200	−2.025436	−0.7566528	−1.167351
205	−2.035645	−0.6841583	12.60843
210	−2.043335	−0.3721161	33.83597
215	−2.044289	0.3075256	66.31113
220	−2.032204	1.548218	115.8063
225	−1.997345	3.642517	190.9517
230	−1.925072	7.028443	304.2704
235	−1.793213	12.35022	472.9959
240	−1.568863	20.52286	718.5657
245	−1.204323	32.76679	1062.373
250	−0.6330795	50.53723	1513.948
255	0.2315712	75.21598	2048.784
260	1.491497	107.454	2582.578

$$\ddot{S}_5 - d\cos\beta \cdot \dot{\beta}^2 - d\sin\beta \cdot \ddot{\beta} = \ddot{x}_D \tag{3.33}$$

$$-d\sin\beta \cdot \dot{\beta}^2 + d\cos\beta \cdot \ddot{\beta} = \ddot{y}_D \tag{3.34}$$

The calculations correspond to the case presented in Fig. 3.9, considering the crank's rotation velocity of 60 rpm. Figure 3.16 presents the diagram of the velocity for slider 5 (Fig. 3.8), that is $S_5'(\varphi)$. One can notice almost zero velocities for the

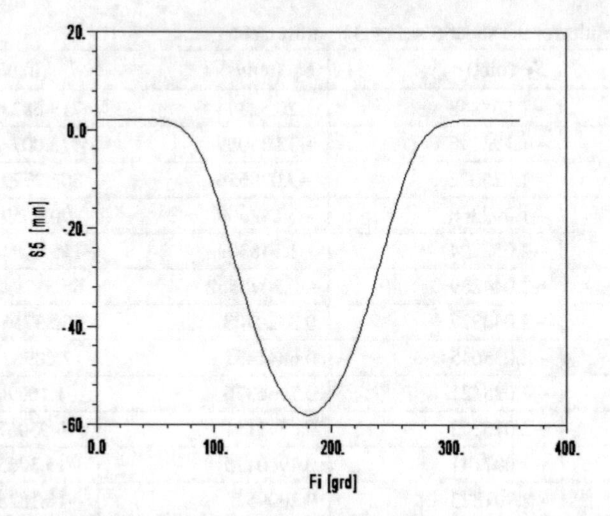

Fig. 3.15 Variation $S_5(\varphi)$ for the arc on the opposite side

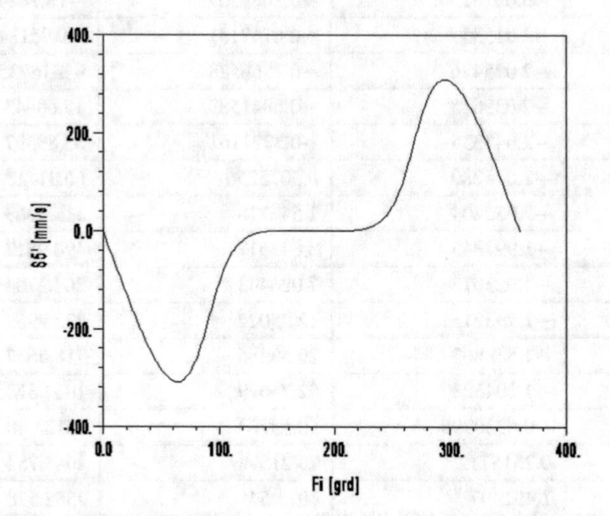

Fig. 3.16 Variation $S_5'(\varphi)$

pause interval indicated in Fig. 3.13. From Table 3.2, one can see that in the area of interest, the velocities are smaller than 1 mm/s, and therefore are acceptable.

Figure 3.17 presents the variations of the velocity and acceleration for the slider 5, for the whole working cycle. In the area of interest, the values are very small (Table 3.2).

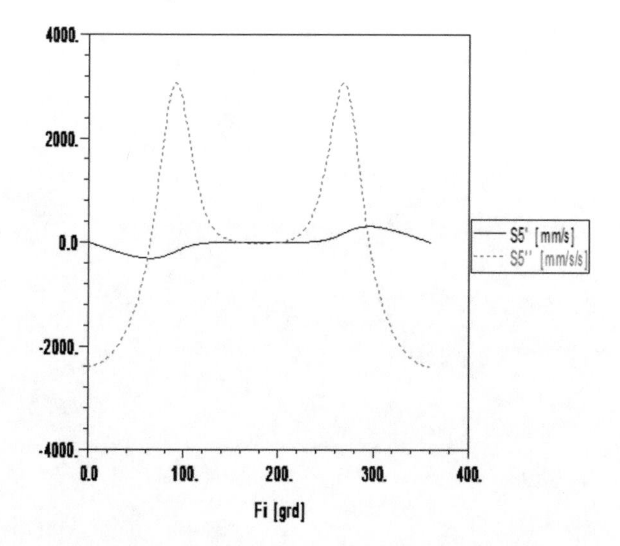

Fig. 3.17 Variation of velocity and acceleration for a whole working cycle

References

1. Artobolevskii II (1971) Mehanizmî v sovremennoi tehnike, vol I–V. Izd. Nauka, Moskva
2. Cebâşev PL (1953) Izobrannâe trud. Izd. Nauka, Moskva
3. Creangă I (1951) ş.a. Curs de geometrie analitică. Editura Tehnică, Bucureşti
4. Popescu I, Luca L, Cherciu M (2013) Structura şi cinematica mecanismelor. Aplicaţii. Editura Sitech, Craiova
5. Popescu I, Luca L, Mitsi S (2011) Geometria, structura şi cinematica unor mecanisme. Editura Sitech, Craiova
6. Reuleaux F (1963) The kinematics of machinery. Dover Publications, New York
7. Reuleaux F, Kennedy ABW (eds) (1876) Kinematics of machinery: outlines of a theory of machines. Macmillan and Co., London
8. Taimina D (2004) Historical mechanisms for drawing curves. Cornell University. https://ecommons.cornell.edu/bitstream/1813/2718/1/2004-9.pdf

Chapter 4
Mechanisms for Generating Conical Curves

Abstract Starting from geometry, the equivalent mechanisms that trace ellipses, hyperbolas, or parabolas are built. These mechanisms are studied structurally, and their kinematic analysis is developed, resulting in relations, successive positions, diagrams, tables, and curves traced. Curves are traced either as trajectories of points or by enveloping. On the basis of projective geometry problems, a new mechanism that traces parabolas by enveloping is presented. Starting from an analytical geometry problem, synthesis of a new equivalent mechanism, which traces parabolas by enveloping, is also created. Next, we study a new mechanism, created on a locus problem, reaching the hyperbolas obtained by enveloping. Another problem in geometry allows the synthesis of the mechanism that traces a hyperbola as a trajectory of a point. A locus problem in 1896 allowed the synthesis of a new mechanism that traces ellipses, parabolas, and hyperbolas.

4.1 Mechanism Ellipsograph

4.1.1 Introduction

Ellipsograph mechanisms have been studied for a long time, being closely linked to the development of the Mechanism and Machine Theory [2, 22]. Scientists, especially mathematicians, have studied various mathematical curves [3, 4, 24]. They have established their geometric properties, and they have drawn it with a ruler and a compass. Afterward, they designed mechanisms to trace them, making models that have verified the correctness of calculations [19]. Taimina studied in detail the evolution of research in the field of the mechanisms generating facet curves [23]. Different mechanisms are described, including conicographs, throughout the evolution of the Theory of Mechanism, from the thirteenth century to the twentieth century. Reuleaux designed and built many mechanisms for curves generating [18, 19]. Artobolevskii also obtained many different mechanisms that draw different mathematical curves, including ellipsographs, parabolographs, and hyperbolographs [1, 2]. All these mechanisms have been designed based on geometrical considerations.

I. Popescu et al., *Mechanisms for Generating Mathematical Curves*,
Springer Tracts in Mechanical Engineering,
https://doi.org/10.1007/978-3-030-42168-7_4

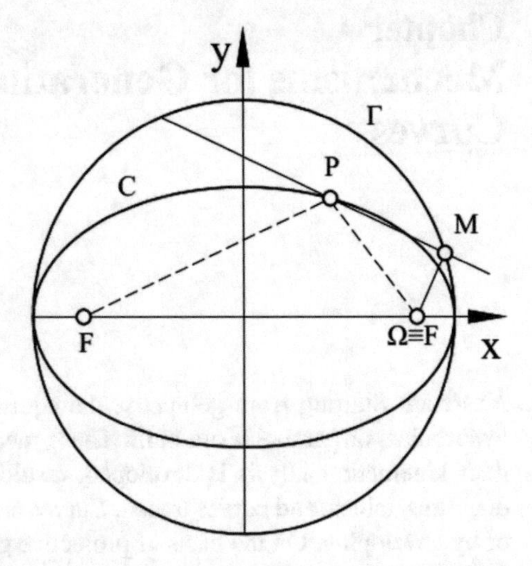

Fig. 4.1 Geometric construction of the pedal Γ of the ellipse C

Figure 4.1 [4] shows the pedal of an ellipse with the focal points F and F'. The pedal curve results from the orthogonal projection of a fixed point on the tangent lines of a given curve.

The locus of point M resulted by bringing a perpendicular from F to the ellipse tangent, in point P. The searched pedal is the Γ circle.

4.1.2 The Mechanism Synthesis

Based on the geometric construction of the pedal (Fig. 4.1), we proposed the inverted problem, that is finding an ellipsograph mechanism [17, 20]. In this way, the point M is taken on the Γ circle, then M is joined to F and further MP is constructed perpendicular on FM, which is the envelope of the ellipse.

Figure 4.2 shows the mechanism which is built based on this method.

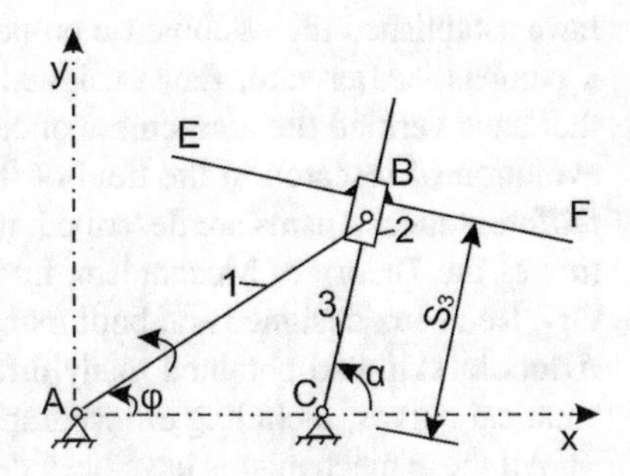

Fig. 4.2 Mechanism obtained

Point C is the focal point of the ellipse. On BC element moves the slider 2 to which the EBF bar is welded, perpendicular to the slide, i.e., on CB. Point B is also on the circle with the radius AB, which is enveloped in ellipse pedals.

4.1.3 The Mechanism Analysis

The mechanism is structurally composed of the driver element AB and the dyad BC, type RPR [8].

Based on Fig. 4.2, Eqs. (4.1)–(4.6) are written:

$$x_B = AB\cos\varphi = x_C + S_3\cos\alpha \tag{4.1}$$

$$y_B = AB\sin\varphi = S_3\sin\alpha \tag{4.2}$$

$$x_F = x_B + BF\cos(\alpha + 270) \tag{4.3}$$

$$y_F = y_B + BF\sin(\alpha + 270) \tag{4.4}$$

$$x_E = x_B + BE\cos(\alpha + 90) \tag{4.5}$$

$$y_E = y_B + BE\sin(\alpha + 90) \tag{4.6}$$

4.1.4 Results Obtained

In the examples in Figs. 4.3 and 4.4, the following numerical data were used: $BF = BE = 80$ mm, X_C and AB are modified on a case by case basis. Initially, we took $AB = 50$ mm and $X_C = 40$ mm.

The mechanism found for a position (Fig. 4.3) which satisfies the imposed geometric conditions.

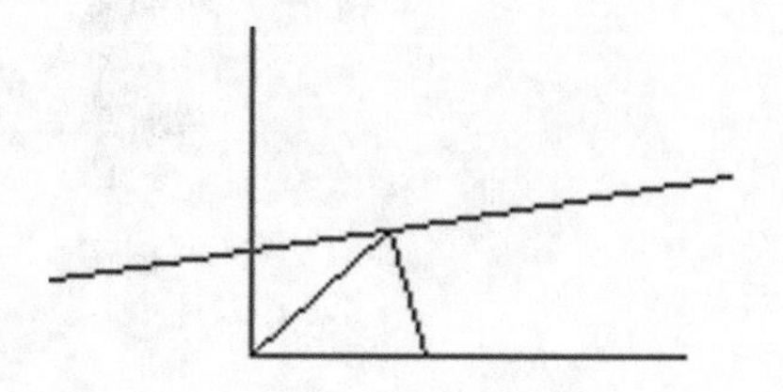

Fig. 4.3 Position of the mechanism

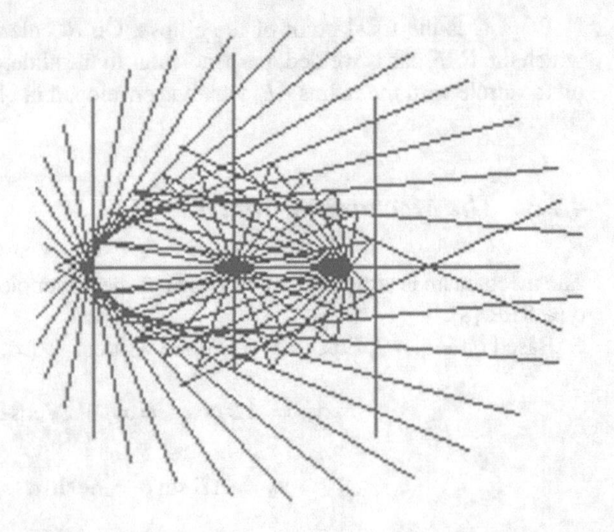

Fig. 4.4 Successive positions of the mechanism

Successive positions of the mechanism are shown in Fig. 4.4, ascertaining the shape of the ellipse generated by enveloping.

The B coordinates variations are given in Fig. 4.5, ascertaining the symmetries specific to the coordinates on a circle [12].

The S_3 trajectory has the curve from Fig. 4.6, free of jumps, so the mechanism works for the entire cycle $\varphi = 0° \ldots 360°$.

The trajectory of point E (Fig. 4.7) is a closed, unusual curve.

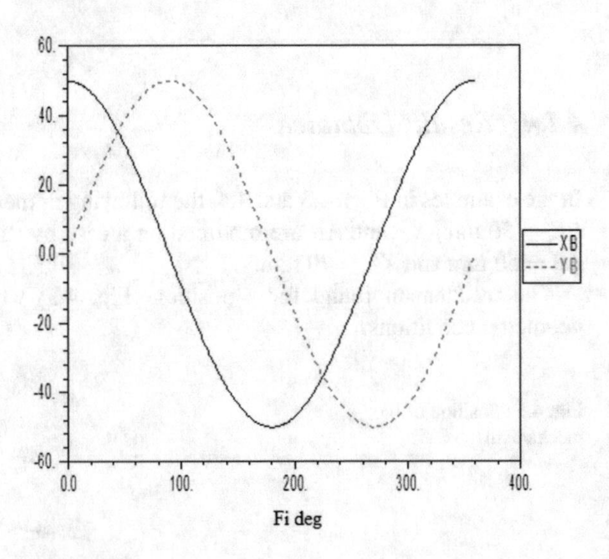

Fig. 4.5 Variations of B coordinates

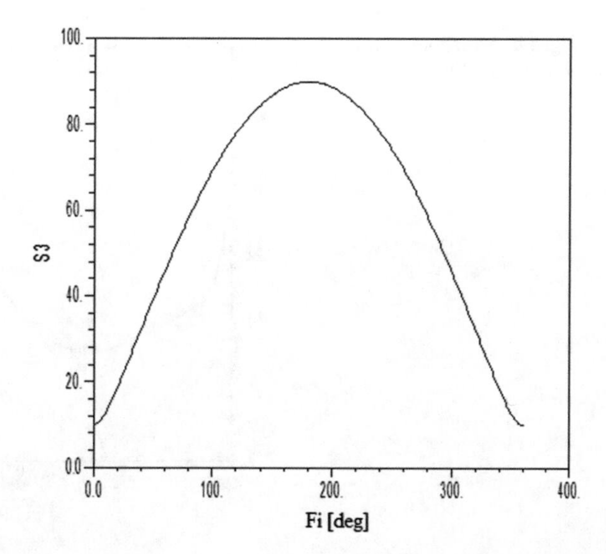

Fig. 4.6 Displacement S_3 (φ)

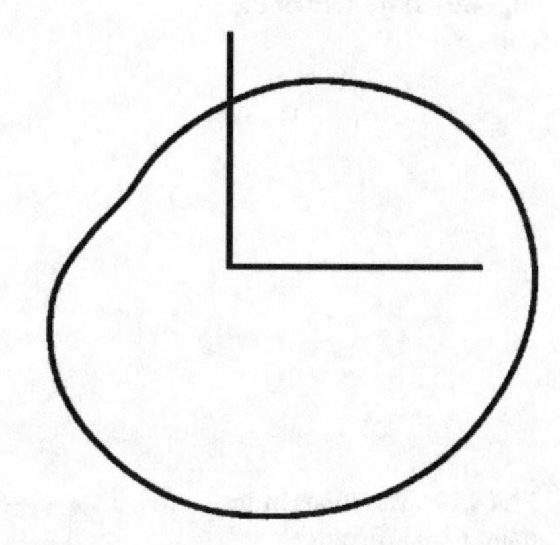

Fig. 4.7 Trajectory of the point *E*

The coordinates of point *E* (Fig. 4.8) are curves without symmetries; it is observed that both curves also reach in the negative area.

Similarly, we studied the motion of the point *F* (Figs. 4.9 and 4.10), also observing the lack of symmetries and the difference from the movement of the point *E*.

Further, for other numerical data: $AB = 100$ mm, $X_C = 40$ mm, the ellipse in Fig. 4.11 results. Compared with Fig. 4.4, a larger and less elongated ellipse can be observed.

Maintaining $AB = 100$ mm, but increasing X_C to the value $X_C = 80$ mm, the ellipse in Fig. 4.12 was obtained, and it is flatter than the one in Fig. 4.11.

If $AB = 100$ mm and $X_C = 0$, we obtained a circle (Fig. 4.13), in this case the focal point being right in the center of the circle.

For $AB = 100$ mm and $X_C = 50$ mm (half of the initial value), the trajectories of points E and F are ellipses type curves (Fig. 4.14).

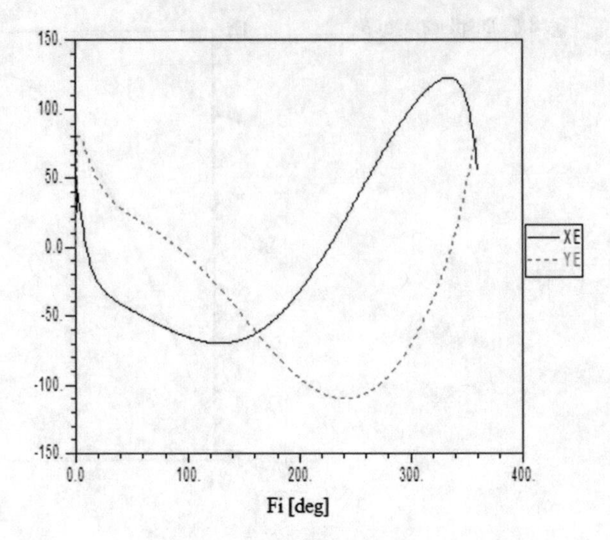

Fig. 4.8 Variations of the point E coordinates

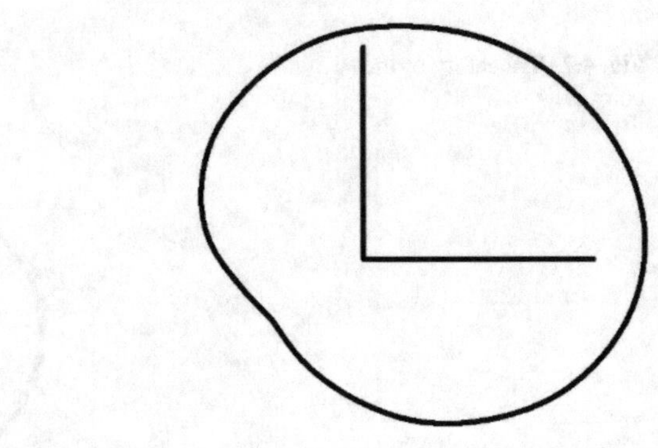

Fig. 4.9 Trajectory of the point F

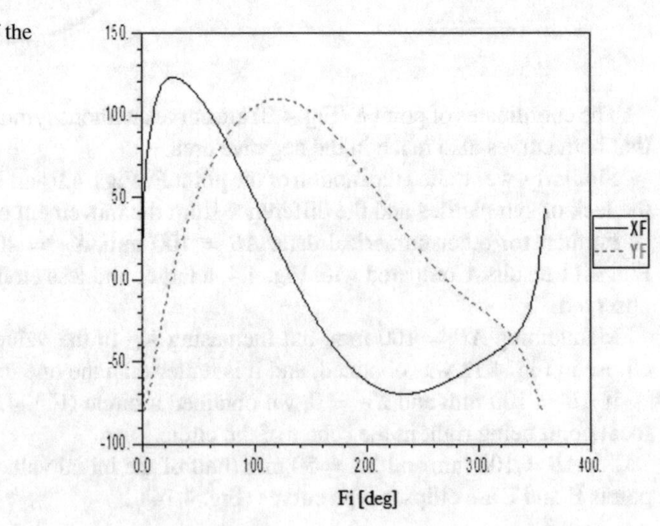

Fig. 4.10 Variations of the point F coordinates

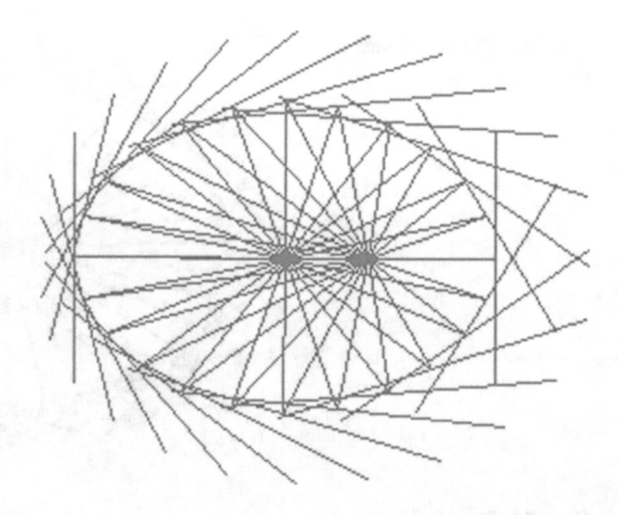

Fig. 4.11 Ellipse with $AB = 100$, $X_C = 40$ mm

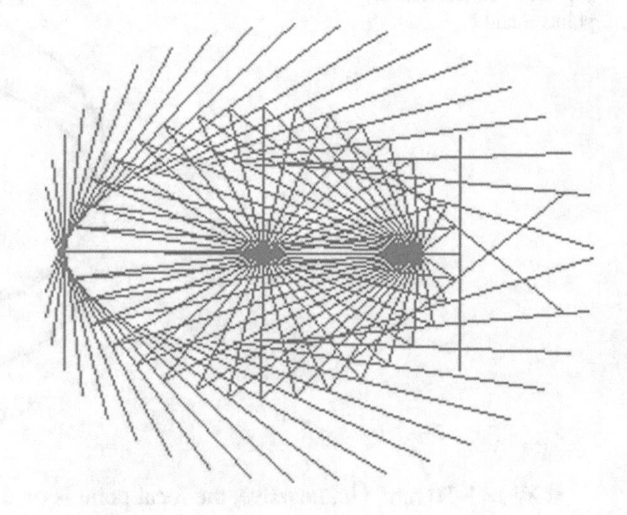

Fig. 4.12 Ellipse with $AB = 100$, $X_C = 80$ mm

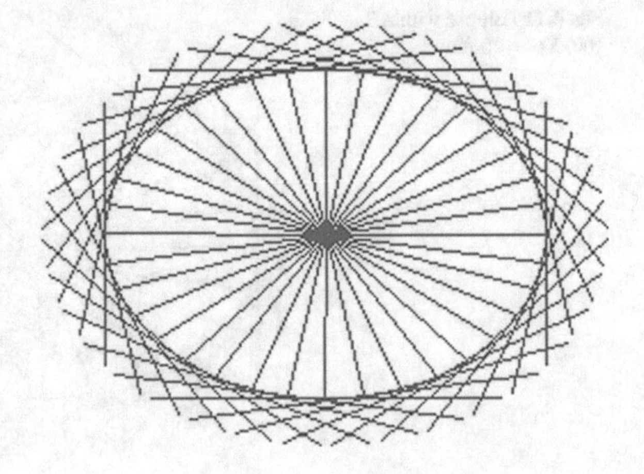

Fig. 4.13 Ellipse becomes a circle

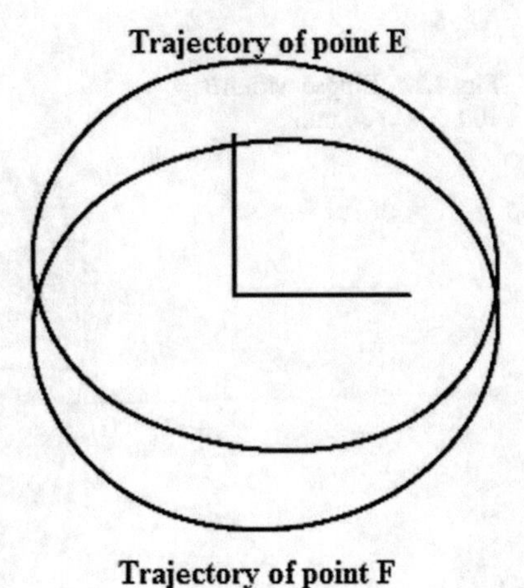

Fig. 4.14 Trajectories of points E and F

If $X_C = -50$ mm, i.e., negative, the focal point is on the left ordinate. For $AB =$ 100 mm, we obtain the ellipse in Fig. 4.15, which is shifted to the left, corresponding to the curves in Fig. 4.16, different from the curves in Fig. 4.10.

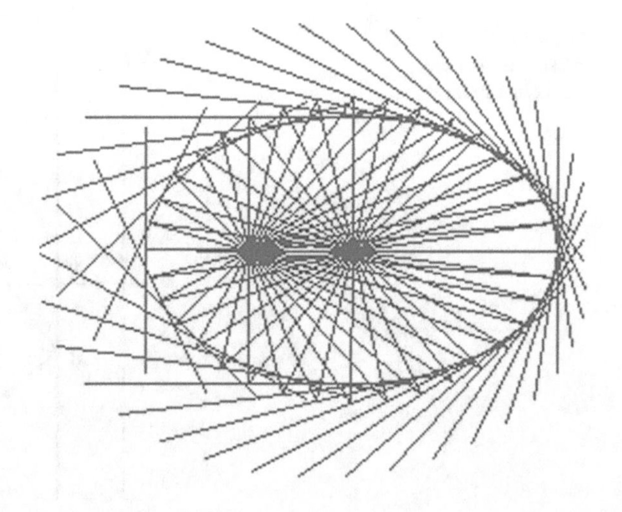

Fig. 4.15 Successive positions

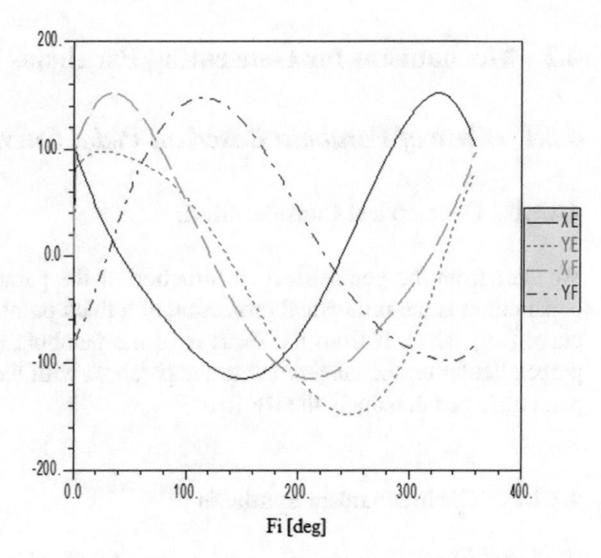

Fig. 4.16 Variations of points E and F coordinates

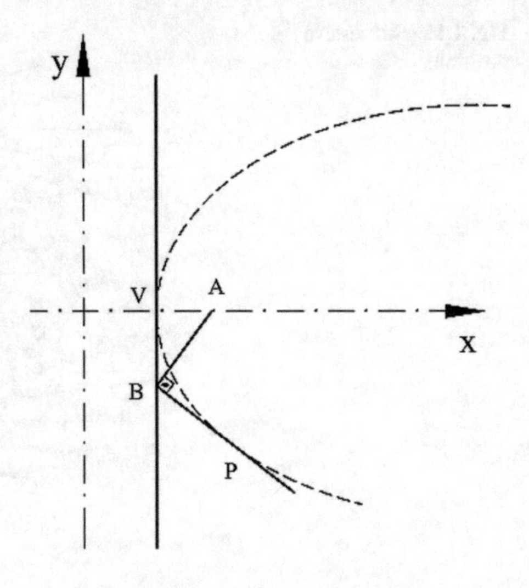

Fig. 4.17 Geometrical construction of the parabola's pedal

4.2 Mechanisms for Generating Parabolas

4.2.1 Case of Parabola Based on Pedal Curve

4.2.1.1 Geometrical Considerations

We start from the geometrical construction of the parabola's pedal, knowing that pedal curve is the orthogonal projection of a fixed point on the tangents of a given curve [24]. Thus, if from the focus *A* of the parabola (Fig. 4.17) we draw *AB*, a perpendicular on the tangent *BP* to the parabola with the vertex *V*, then we get the parabola's pedal, namely the *BV* line.

4.2.1.2 The Mechanism Synthesis

We had the idea about the inverted problem [5, 7]: If we take the directrix line *BV* and the focus point *A*, then the perpendicular *BP* to *AB* will envelop a parabola.

We selected the *CB* line as a directrix (Fig. 4.18) and the *AB* line and the slider 2, which has a point *B* moving on *CB*. The perpendicular *EF* on *AB* will envelop a parabola [8].

For the point *B* to move on the line *CB*, there is the slider 3, and for the length of *AB* to be variable, we have also put the slider 2, which is integral with the line *EF* that envelopes the parabola.

Fig. 4.18 Obtained mechanism

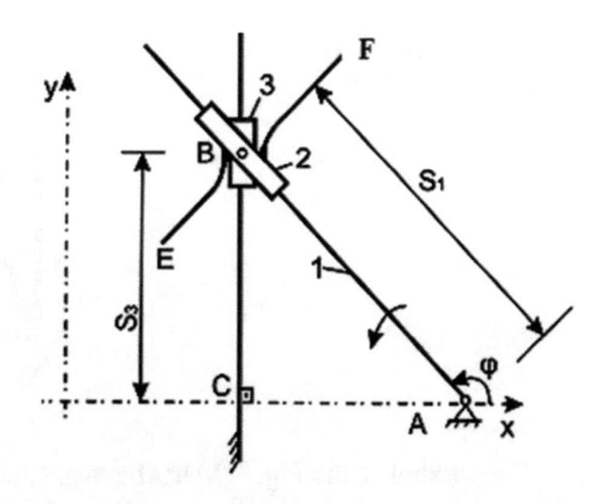

4.2.1.3 The Mechanism Analysis

Structurally, the mechanism is of the R-TRT type [8].

Based on Fig. 4.18, we write Eqs. (4.7)–(4.14):

$$x_B = x_A + S_1 \cos\varphi = x_C = \text{const.} \tag{4.7}$$

$$y_B = y_A + S_1 \sin\varphi = S_3 \tag{4.8}$$

$$x_F = x_B + BF\cos(\varphi - 90) \tag{4.9}$$

$$y_F = y_B + BF\sin(\varphi - 90) \tag{4.10}$$

$$x_E = x_B + BE\cos(\varphi + 90) \tag{4.11}$$

$$y_E = y_B + BE\sin(\varphi + 90) \tag{4.12}$$

$$S_1 = (x_C - x_A)/\cos\varphi \tag{4.13}$$

$$S_3 = y_A + S_1 \sin\varphi \tag{4.14}$$

4.2.1.4 Results Obtained

Figure 4.19 shows the obtained mechanism for $\varphi = 120°$. We can observe the vertical line CB (moved with X_C to the right side of the axis system's ordinate), the variable line *AB*, and the tangent *EF* to the parabola.

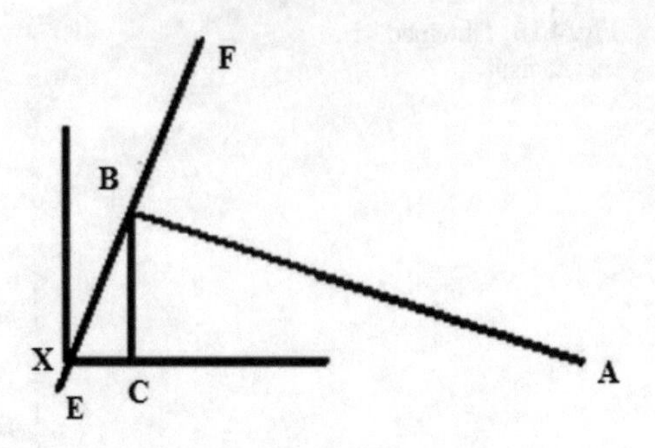

Fig. 4.19 Mechanism in the position for $\varphi = 120°$

The parabola from Fig. 4.20 has been generated, with $X_A = 100$ mm, $X_C = 25$ mm. We can observe the successive positions of the tangent and the shape of the parabola, as a result.

Next, we keep $X_A = 200$ mm, but the distance X_C has been modified, which is the position of the CB line. For $X_C = 10$ mm, we obtain Fig. 4.21, similar to Fig. 4.20, but with a wider span of the parabola (the length between the extreme points).

The parabola from Fig. 4.22 ($X_C = 0$) is different from the one in Fig. 4.23 ($X_C = 60$ mm).

Figure 4.24 depicts the parabola with $X_C = 150$ mm.

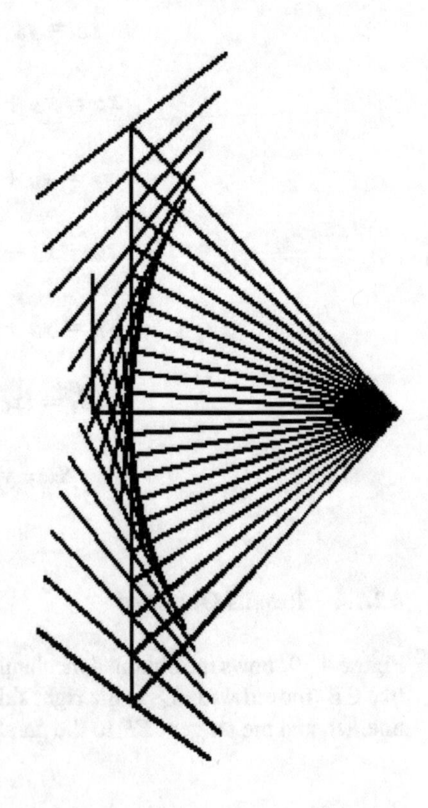

Fig. 4.20 Successive positions of the tangent

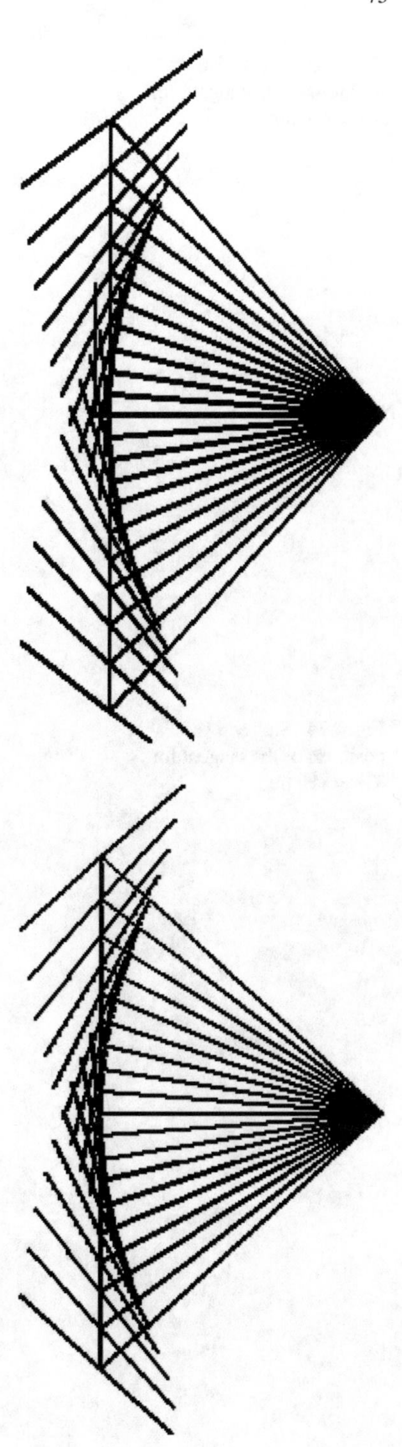

Fig. 4.21 Successive positions of the tangent for $X_C = 10$ mm

Fig. 4.22 Successive positions of the tangent for $X_C = 0$

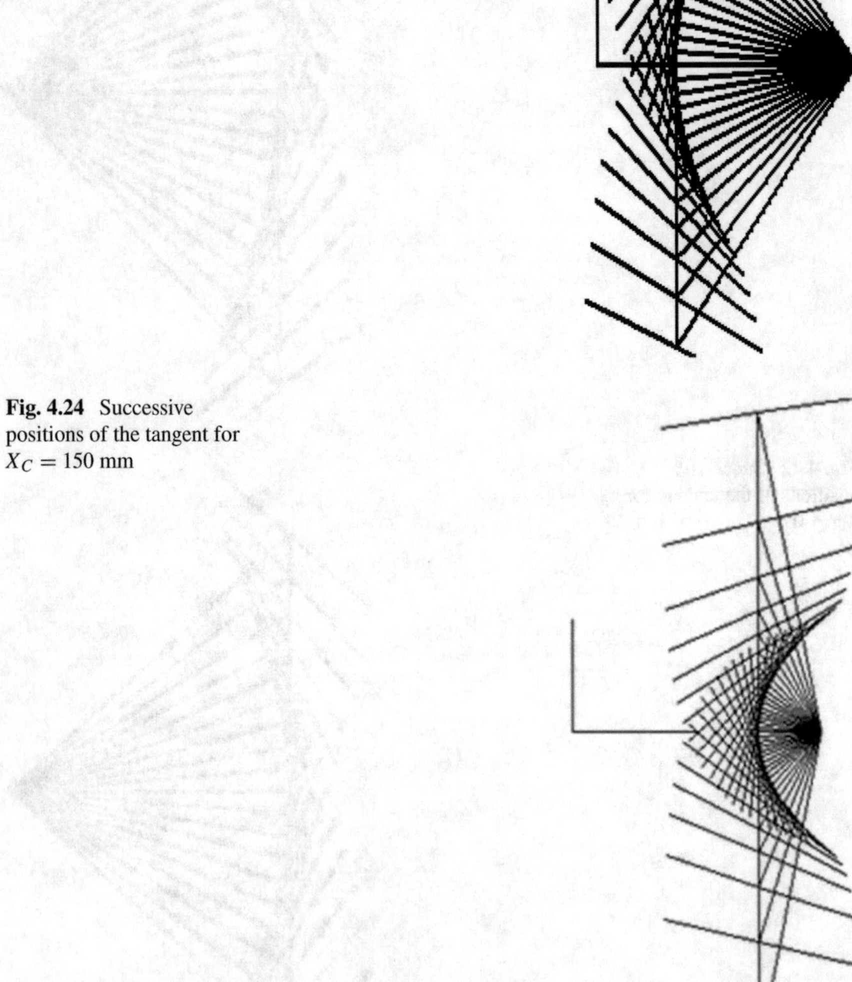

Fig. 4.23 Successive positions of the tangent for $X_C = 60$ mm

Fig. 4.24 Successive positions of the tangent for $X_C = 150$ mm

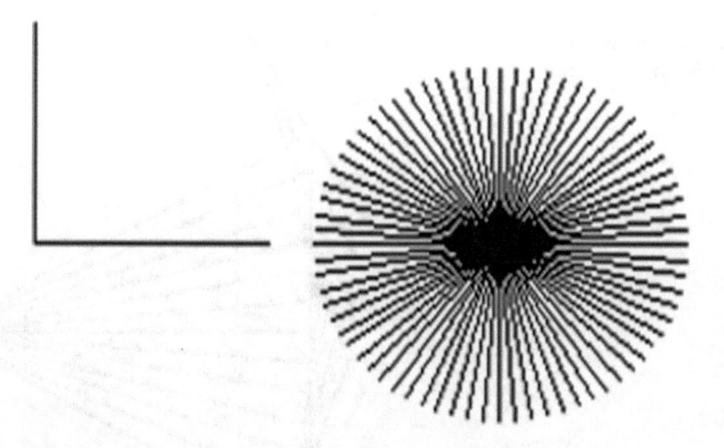

Fig. 4.25 Successive positions of the tangent for $X_C = X_A = 200$ mm

Figure 4.25 represents the parabola for $X_C = X_A = 200$ mm, which means a circle, if the line *CB* goes through the focus point *A*.

The parabola in Fig. 4.26 has $X_C = -25$ mm, meaning the CB line is positioned to the left side of the ordinate axis, and the one in Fig. 4.27 has $X_C = -50$ mm, the difference between their distances being obvious.

For the case $X_A = 60$ mm and $X_C = 25$ mm, meaning X_A is smaller than previous cases, the resulted parabola (Fig. 4.28) has a different shape (narrower) than the

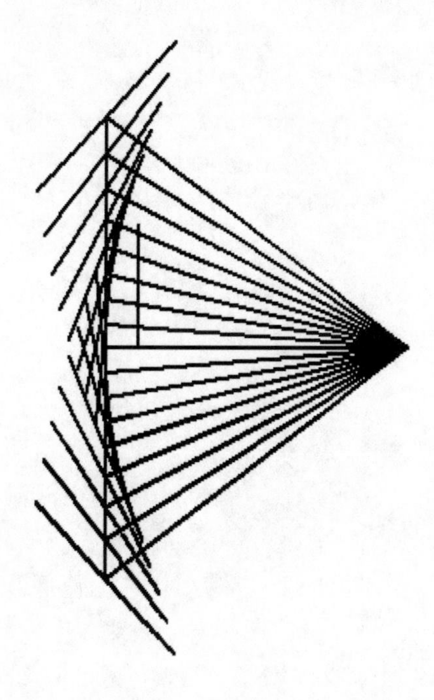

Fig. 4.26 Successive positions of the tangent, for $X_C = -25$ mm

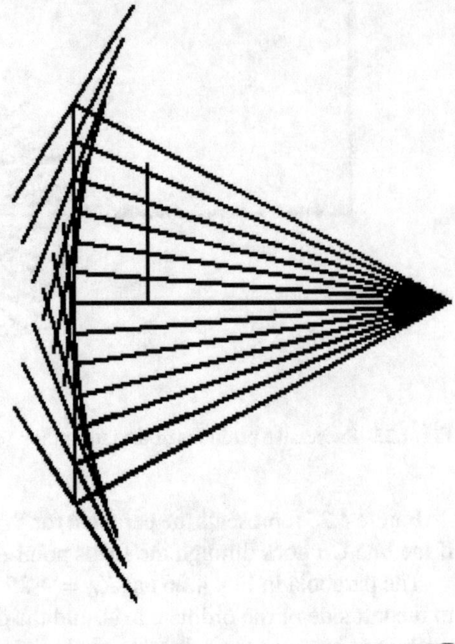

Fig. 4.27 Successive positions for $X_C = -50$ mm

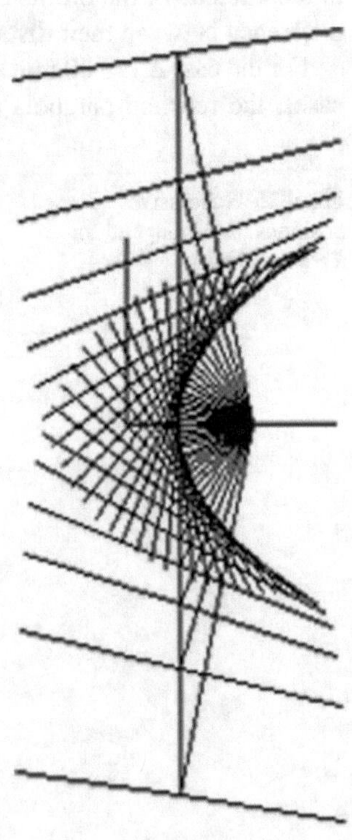

Fig. 4.28 Successive positions for $X_A = 60$ mm, $X_C = 25$ mm

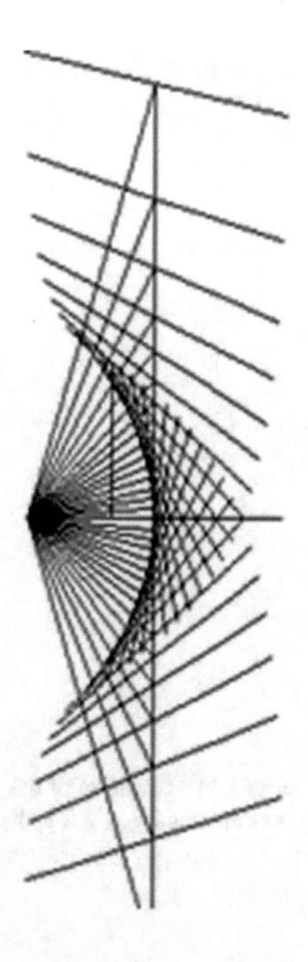

Fig. 4.29 Successive positions for $X_A = -50$ mm, $X_C = 25$ mm

previous ones (Figs. 4.21, 4.22, 4.23, 4.24, 4.25, 4.26 and 4.27) and also different dimensions.

For $X_A = -50$ mm and $X_C = 25$ mm, the parabola in Fig. 4.29 is positioned in reverse compared to the one in Fig. 4.28 because the position of the focus point has changed.

For the above values, $X_A = 200$ mm, $X_C = 25$ mm, the trajectories of the extreme points of the tangent *EF* are given in Fig. 4.30.

From the diagram in Fig. 4.31 and Table 4.1, we can see the similarities between the drawn curves by *E* and *F*, on some areas. We have the following values: $X_A = 200$, $X_C = 25$, $BE = BF = 80$ mm.

The version with $Y_A > 0$ was also tried, resulting in Fig. 4.32, the parabola that is the same as that of Fig. 4.20 (other initial data are the same), but shifted by Y_A above to abscissa.

Referring to Fig. 4.31 and Table 4.1, it is observed that the mechanism does not work for two subintervals: $\varphi = 60° \ldots 120°$ and $\varphi = 240° \ldots 300°$. This is the reason for which the radius *AB* does not reach all trigonometric quadrants in previous figures of parabolas. The explanation is that at certain values of φ, point *B* tends to be infinite on the fixed line *CB* (the program provided instructions to jump over large values).

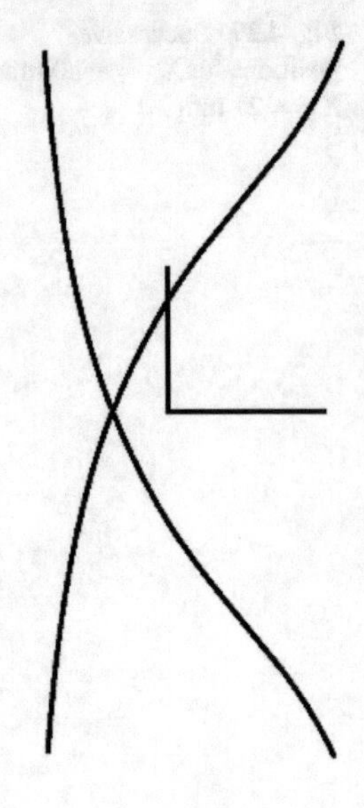

Fig. 4.30 Trajectories of E and F

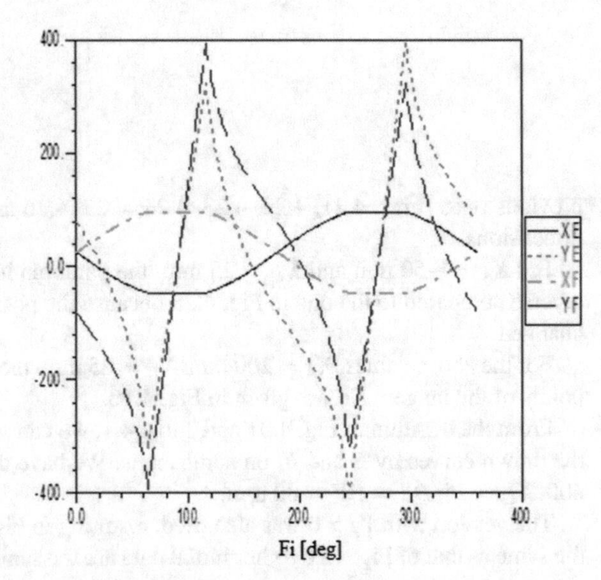

Fig. 4.31 Variation of the coordinates X_E, Y_E X_F, Y_F

4.2.2 A New Mechanism for Generating Parabola

4.2.2.1 Geometrical Considerations

In projective geometry [3], the curves whose points are equally spaced from their directrix are called symmetrically. In the case of the parabola (Fig. 4.33), there is considered to be a point F and a straight line d.

Table 4.1 Values for coordinates of points E and F

F_i	X_E	Y_E	X_F	Y_F
0	25.00009	80	25.00009	−80
10	11.10825	47.92745	38.89193	−109.6418
20	−2.361511	11.4807	52.36169	−138.8701
30	−14.99989	−31.7541	65.00005	−170.3182
40	−26.42291	−85.55864	76.42305	−208.1258
50	−36.28346	−157.1335	86.28358	−259.9796
60	−44.28196	−263.1082	94.28204	−343.1083
120	−44.28215	263.1102	94.28206	343.11
130	−36.28369	157.1347	86.28358	259.9806
140	−26.42321	85.55962	76.42305	208.1265
150	−15.00022	31.75483	65.00006	170.3187
160	−2.361858	−11.48013	52.36168	138.8706
170	11.10789	−47.92695	38.89195	109.6422
180	24.99973	−79.9996	25.00009	80.00041
190	38.89155	−109.6414	11.10826	47.92789
200	52.36132	−138.8698	−2.361504	11.48117
210	64.99973	−170.3178	−14.99987	−31.75353
220	76.42278	−208.1253	−26.4229	−85.55791
230	86.28335	−259.9788	−36.28344	−157.1324
240	94.28185	−343.107	−44.28196	−263.1065
300	94.28223	343.1113	−44.28215	263.1118
310	86.28383	259.9812	−36.28369	157.1357
320	76.42333	208.1269	−26.42318	85.56021
330	65.0004	170.3192	−15.00022	31.75544
340	52.36205	138.8709	−2.361858	−11.47963
350	38.89231	109.6426	11.10789	−47.92654
360	25.00047	80.0008	24.99973	−79.9992

The distances from points B, B' to F and d are r, r'. Successive positions of B describe a parabola, if the radius r, r', r'', etc. will be larger than FS, where S is the middle point of the perpendicular from F on d.

4.2.2.2 Synthesis of the Equivalent Mechanism

Based on the geometrical considerations, we created the mechanism presented in Fig. 4.34 [11]. The driving link 1 has a translational movement through the slider A which connects 1 to the fixed base (0).

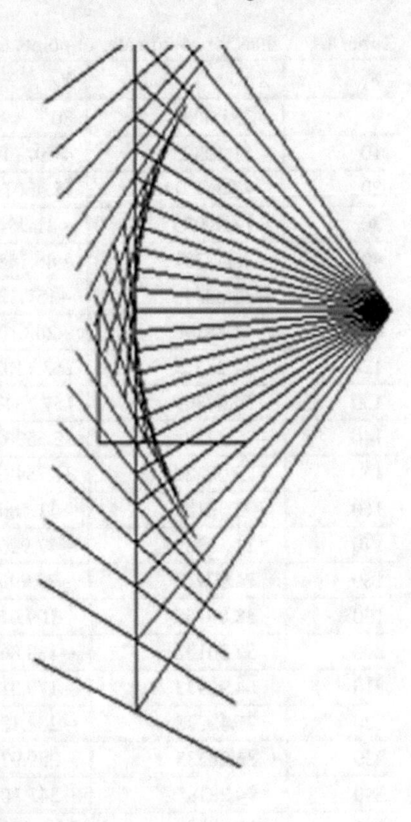

Fig. 4.32 Parabola for $Y_A > 0$

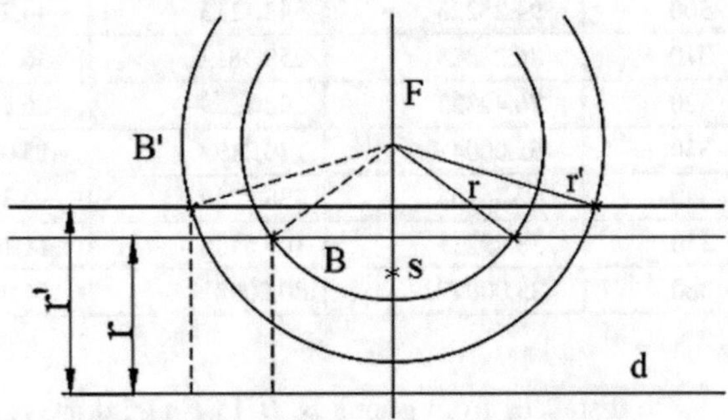

Fig. 4.33 Principle of parabola generation

Because the distance *AB* must be adjustable, the slider 1–2 of *B* was provided. The point *C* was taken as a fixed point, and the radius variation was ensured through the slider 5–3 at *B*.

The difficult problem, solved after several attempts, was to ensure that *B* is equal spaced to *C* and *xx*. The solution was to construct a determined isosceles triangle *ABC*.

For this, the line *DB* was imposed as the height of the triangle; the angle at *D* was assigned to 90°, and the slider 6–4 was provided at *B*.

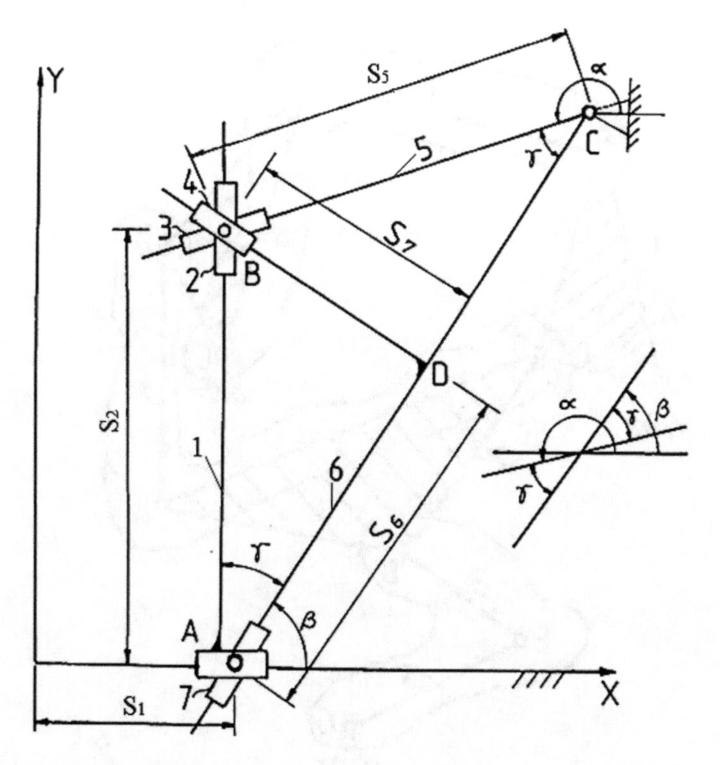

Fig. 4.34 Resulted mechanism

This ensures for *B* to be a fixed perpendicular to *AC*, and at the successive positions of *A* or of *B*, the straight line *AC* which is perpendicular to *BD* will exist.

To achieve the drawing in Fig. 4.34, first *C* was adopted, then *B* and the radius *BC* was given to build a circle; the x-axis was adopted at a point of tangency. Thus, *BA* and *BC* are always the radius of the circle with center in *B*.

From the semicircle subtended by *AB* (extended), it is observed that the angle *ABC* is equal to the arc *AC*, and the angle *BAC* is equal to (180-*ABC*)/2.

The semicircle *CA*, generated by the diameter 5 (*CB* extended), results that the *CBA* arc is equal to the *CBA* angle, the rest being comprised by the *BCA* angle = (180-*CBA*)/2. Thus, if the γ angles are equal, then the triangle is isosceles.

The structure of the mechanism, given in Fig. 4.35, appears to be a mechanism with three dyads, type: P-RPR-PRP-RPR [8].

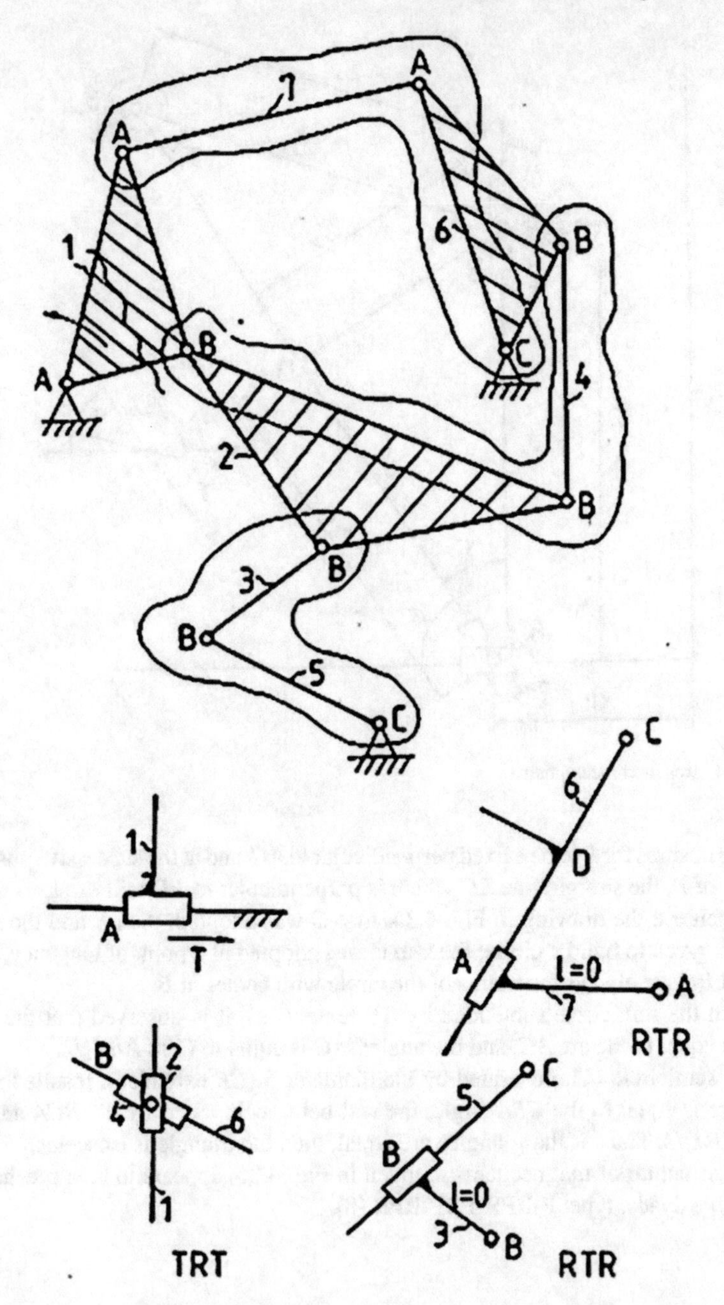

Fig. 4.35 Structure of the mechanism

4.2.2.3 The Mechanism Analysis

To establish the kinematic possibilities of the mechanism in Fig. 4.34, Eqs. (4.15, 4.16) are written:

From:

$$\tan\beta = y_C \big/ (x_C - S_1) \tag{4.15}$$

resulting β and of:

$$\gamma = 90 - \beta \tag{4.16}$$

resulting γ.

The angle α is then calculated with (4.17) (as detailed in Fig. 4.34):

$$\alpha - \beta + \gamma = 180 \tag{4.17}$$

Through the method of contours [8] results in Eqs. (4.18, 4.19):

$$x_B = x_C + S_5 \cos\alpha \tag{4.18}$$

$$y_B = S_2 = y_C + S_5 \sin\alpha \tag{4.19}$$

where S_5 and S_2 are calculated.

Next, S_7 is obtained from (4.20) and S_6 from (4.21):

$$\sin\gamma = S_7/S_5 \tag{4.20}$$

$$\tan\gamma = S_7/S_6 \tag{4.21}$$

4.2.2.4 Results Obtained

Based on Eqs. (4.15)–(4.21), a computer program has been achieved, for both tabular and graphical results that have been obtained [12].

The tables revealed the following: The values of S_2 and S_5 (Fig. 4.34) are equal to any position of the mechanism, so it provides equal distances to C and xx. The values of the radius r, r' (Fig. 4.33) are always greater than FS. In the case of Fig. 4.34, it means that $S_5 = S_2 > Y_C/2$, as is respected in the tabulated results.

When Y_C increases, the curve moves above and thereby increasing curvature, and if the curve Y_C decreases, then the curve descends and the curvature decreases (Fig. 4.36).

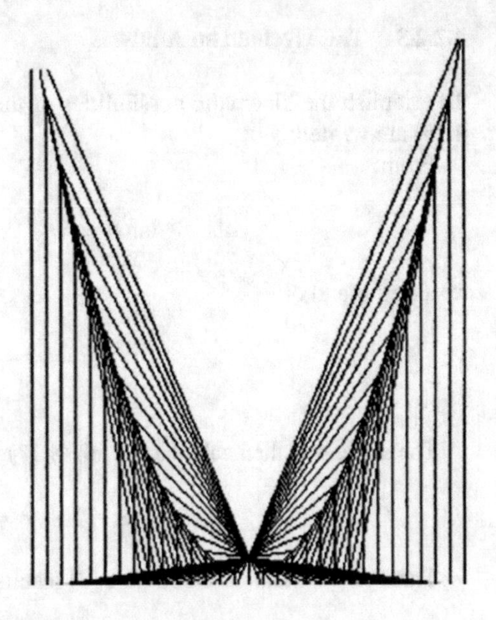

Fig. 4.36 Successive positions of the mechanism for $X_C = 50$ mm, $Y_C = 5$ mm

In Fig. 4.37, the successive positions of the mechanism are shown, where it is clearly observed the drawn parabola, for $Y_C = 45$ mm.

If the focus is positioned on the *y*-axis, the curve is symmetrical with this axis.

If $Y_C < 0$, the curve is moving below the *x*-axis and the parabola is inverted as orientation (Fig. 4.38).

Other conditions are given in Figs. (4.39, 4.40, 4.41, 4.42 and 4.43).

Referring to Fig. 4.39, if the coordinates of *C* are equal, the left branch of the parabola is shorter. In Fig. 4.40, it can also be noted a parabola with unequal branches.

The parabola in Fig. 4.41, having equal branches, was obtained for $X_C = 50 > Y_C = 30$ mm.

The case from Fig. 4.42 corresponds to $X_C = 0 < Y_C = 80$ mm. It is noted the left branch of the parabola being very short.

The parabola in Fig. 4.43 was obtained for $X_C = Y_C = 60$ mm having the right branch shorter.

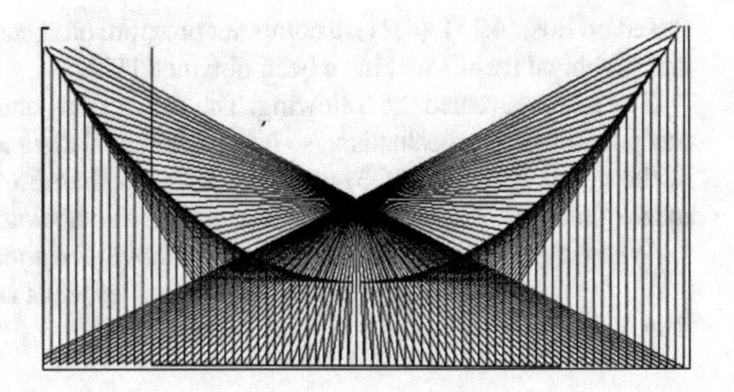

Fig. 4.37 Successive positions of the mechanism for $X_C = 50$ mm, $Y_C = 45$ mm

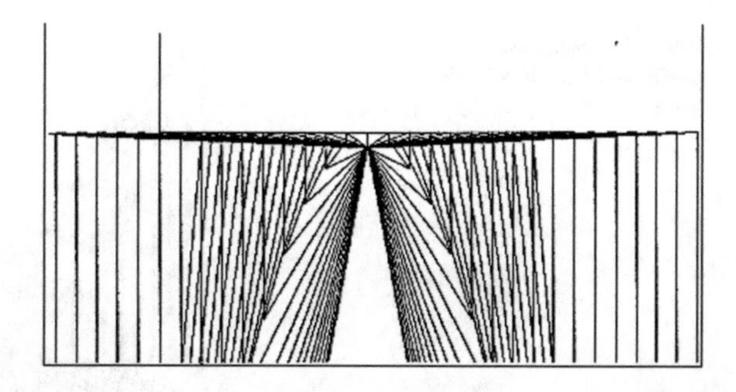

Fig. 4.38 Successive positions of the mechanism for $X_C = 50$ mm, $Y_C = -5$ mm

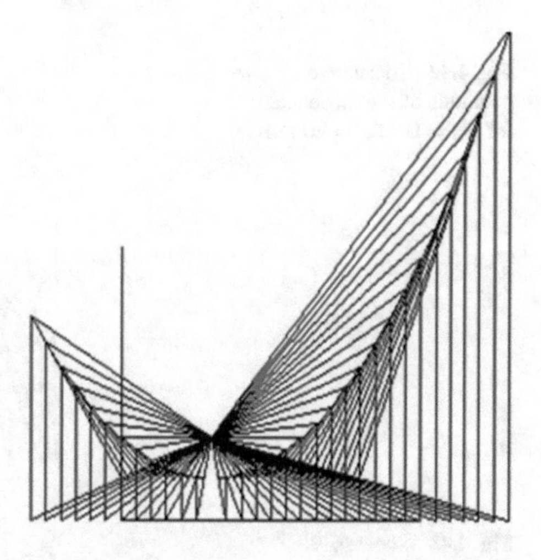

Fig. 4.39 Successive positions of the mechanism for $X_C = Y_C = 30$ mm

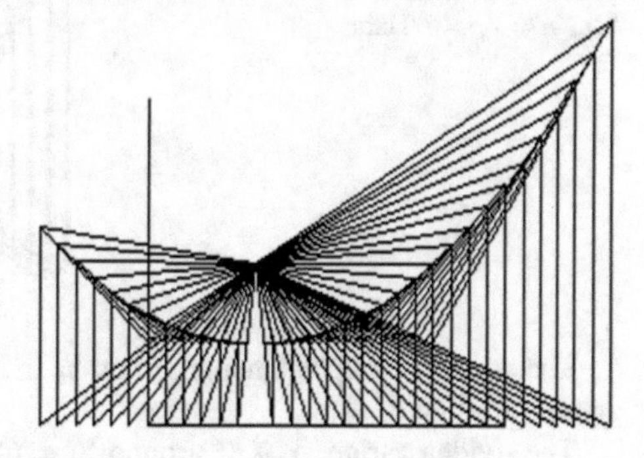

Fig. 4.40 Successive positions of the mechanism for $X_C = 30 < Y_C = 50$ mm

Diagrams

In Fig. 4.44, the variations of coordinates of points *B* and *D* are given for $X_C = Y_C = 60$ mm. The variations are linear except the curve of Y_B. It is interesting that Y_D results as being constant, although the point *D* is moving.

Fig. 4.41 Successive positions of the mechanism for $X_C = 50 > Y_C = 30$ mm

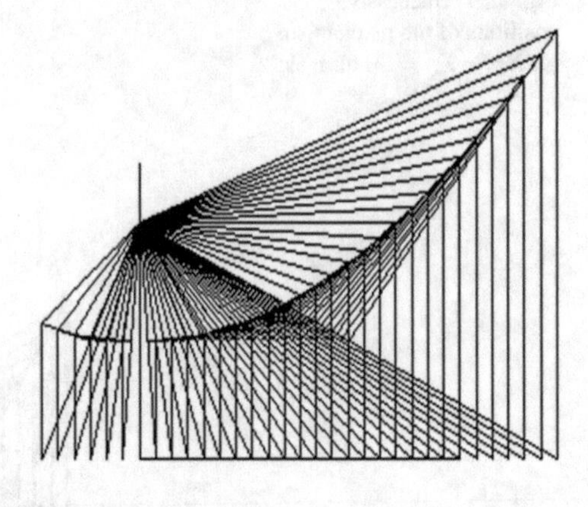

Fig. 4.42 Successive positions of the mechanism for $X_C = 0 < Y_C = 80$ mm

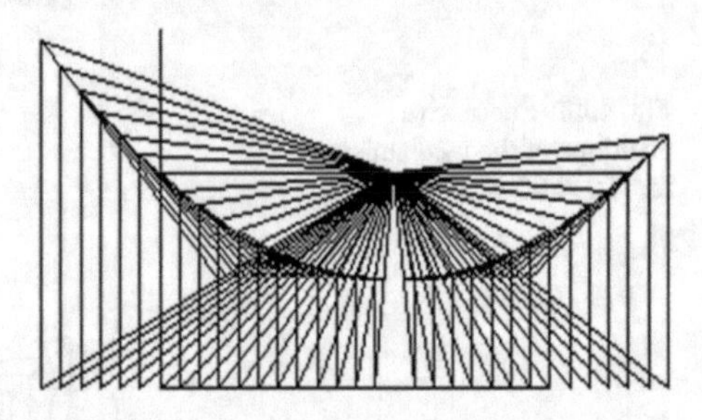

Fig. 4.43 Successive positions of the mechanism for $X_C = Y_C = 60$ mm

The curves of displacements S_2 and S_6 in Fig. 4.45 are nonlinear in terms of the linear variation of S_1.

The sudden variation of S_6 around $S_1 = 60$ mm is due to a jump of the angle β with the changing of trigonometric quadrant as shown in Fig. 4.46. The other angles have that jump too, when reaching the 270° value, because of the arctan function.

In Fig. 4.47, the curves for displacements S_5 and S_7 are given, both variations being nonlinear.

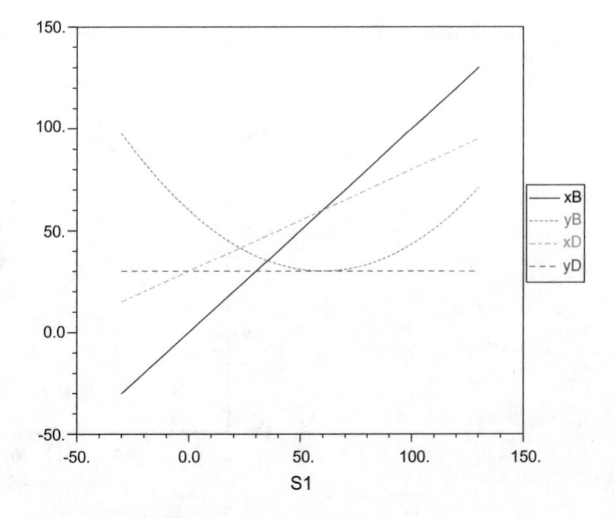

Fig. 4.44 Variations of points B and D coordinates

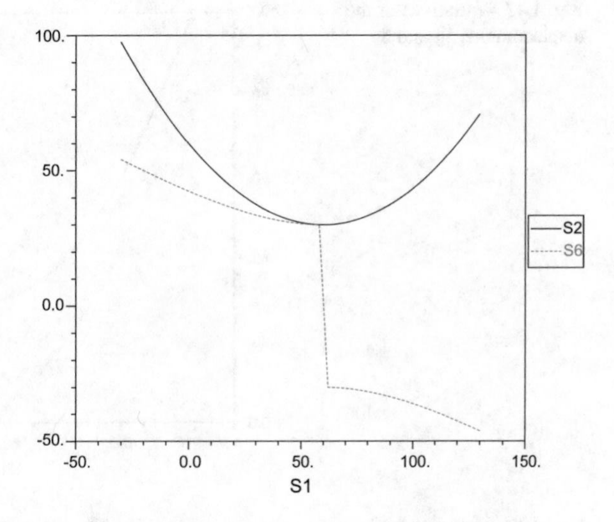

Fig. 4.45 Displacements S_2 and S_6

4.2.3 Case for Generation of Parabola as Envelope of a Line

4.2.3.1 Geometrical Considerations

It starts from a method of generating a parabola as an envelope of a line, given in the projective geometry [3]. It is shown that if we take the focus F (Fig. 4.48), a tangent TT and the segments FM_i are going from F, choosing the points M_i on the tangent TT, then perpendiculars to FM_i, in M_i, will generate a parabola as their envelope.

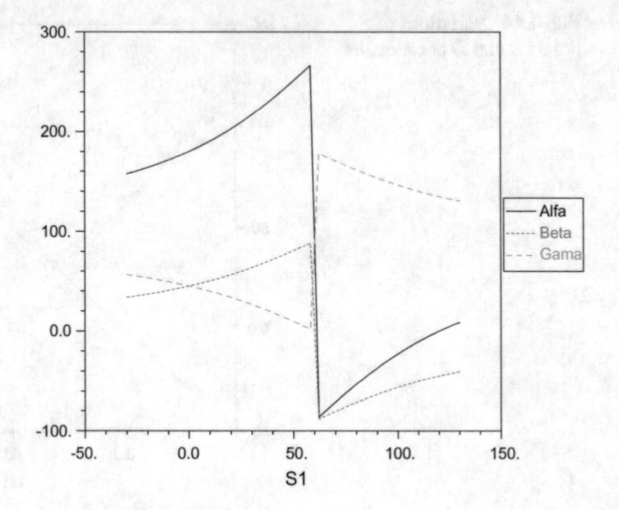

Fig. 4.46 Variations of the α, β, γ angles

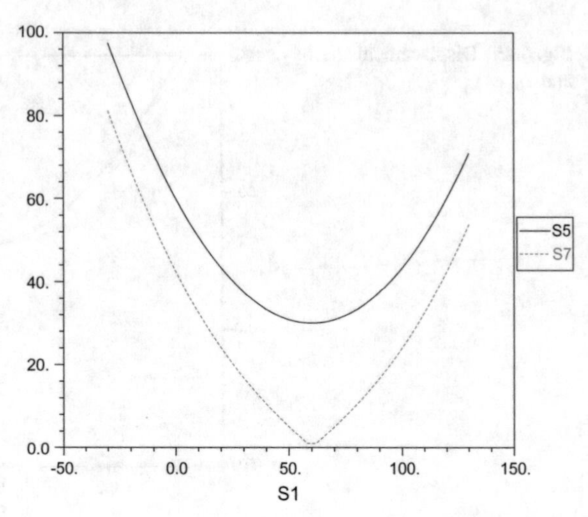

Fig. 4.47 Variations of the displacements S_5 and S_7

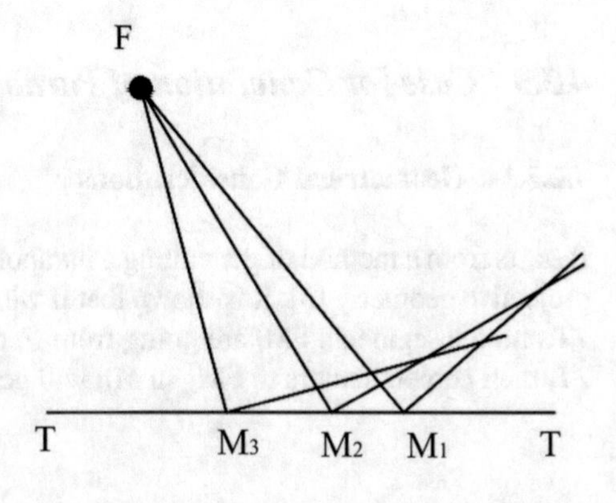

Fig. 4.48 Geometrical construction of the parabola as envelope of a line

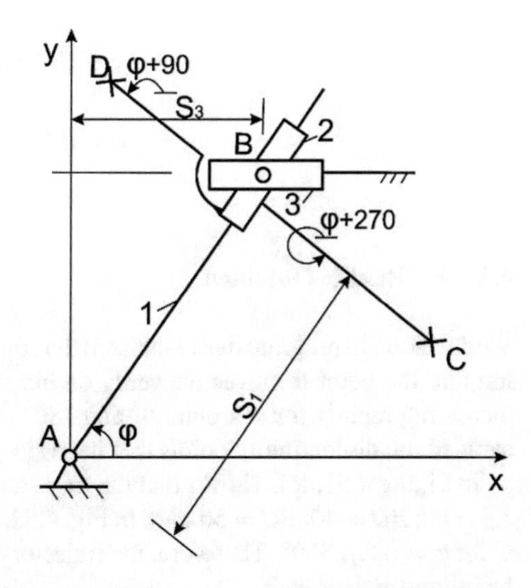

Fig. 4.49 Mechanism for generation of parabola as envelope of a line

4.2.3.2 The Mechanism Synthesis

Referring to Fig. 4.48, the following are shown: FM_i radii have variable length; M_i points are always on *TT*; the perpendicular to FM_i passes through M_i, that is the common point of the lines FM and *TT*.

Based on these considerations, it was designed the mechanism in Fig. 4.49 [9], wherein the driving link has been taken as the element 1, and the line that generates the curve, *CD*, is connected to the slider 2, at point *B*, being perpendicular to *AB*.

4.2.3.3 The Mechanism Analysis

On the basis of Fig. 4.49, we write Eqs. (4.22)–(4.29):

$$x_B = S_1 \cos\varphi = S_3 \tag{4.22}$$

$$y_B = S_1 \sin\varphi = \text{constant} \tag{4.23}$$

$$S_1 = y_B / \sin\varphi \tag{4.24}$$

$$S_3 = S_1 \cos\varphi \tag{4.25}$$

$$x_C = x_B + BC\cos(\varphi + 270) \tag{4.26}$$

$$y_C = y_B + BC\sin(\varphi + 270) \tag{4.27}$$

$$x_D = x_B + BD\cos(\varphi + 90) \tag{4.28}$$

$$y_D = y_B + BD\sin(\varphi + 90). \tag{4.29}$$

4.2.3.4 Results Obtained

We developed a program that calculated the coordinates of point B (Fig. 4.49), observing that the point B moves unevenly on his axis (parallel to xx), the values of X_B increasing rapidly for φ around 0° and 180° (Fig. 4.50) (therefore, there have been avoided the discontinuity points that have sin φ as denominator).

In Figure 4.51, it is shown that the trajectory of point C, for $\varphi = 0° \dots 180°$ and $Y_B = 80$; $BD = 40$; $BC = 50$ mm. In Fig. 4.52, it is shown that the trajectory of point C for $\varphi = 0° \dots 360°$. Therefore, the trajectory of C has two branches which tend to be infinite at their ends.

The trajectory of point D is similar to that of point C (Fig. 4.53), having also two branches, but not identical as it is shown in Fig. 4.54.

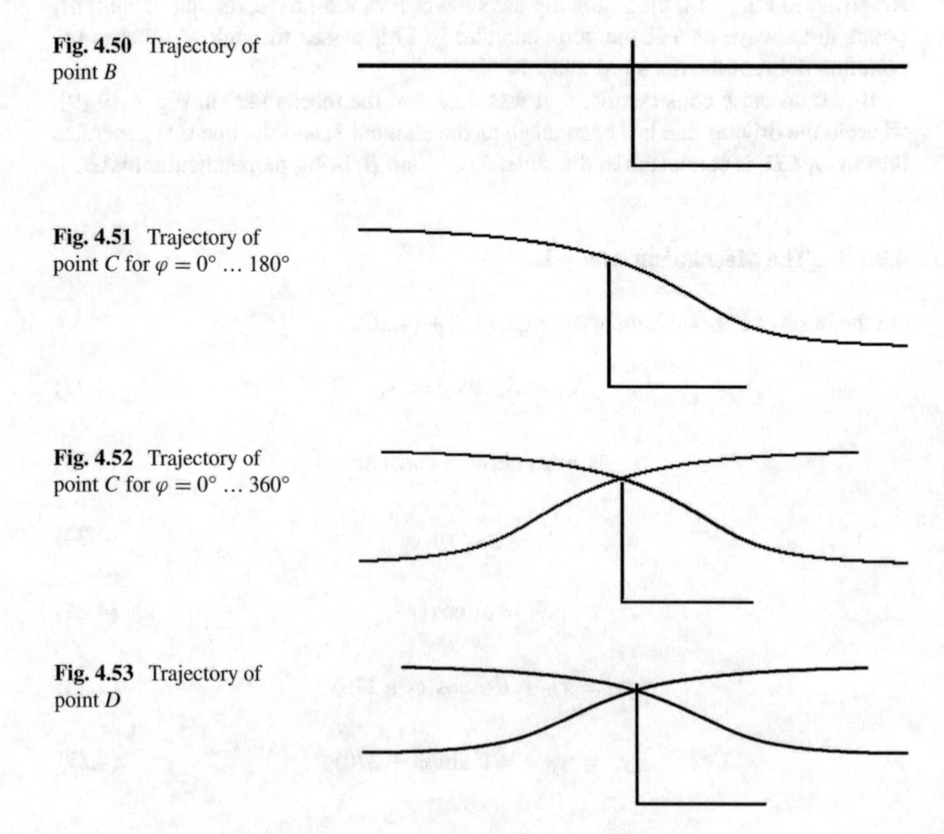

Fig. 4.50 Trajectory of point B

Fig. 4.51 Trajectory of point C for $\varphi = 0° \dots 180°$

Fig. 4.52 Trajectory of point C for $\varphi = 0° \dots 360°$

Fig. 4.53 Trajectory of point D

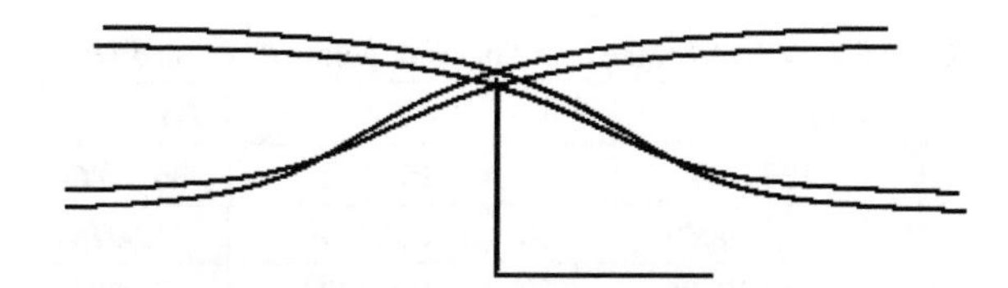

Fig. 4.54 Trajectories of points *C* and *D*

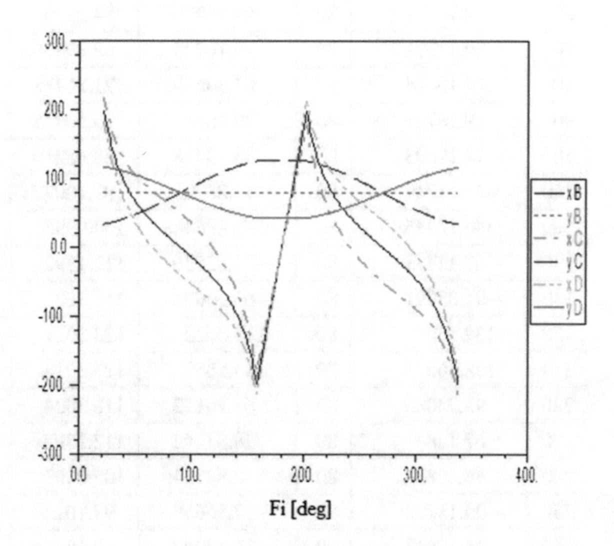

Fig. 4.55 Coordinates of points *B*, *C*, and *D*

In Fig. 4.55, it is shown the variation of the coordinates of points *B*, *C*, *D* with φ. Due to the infinite jumps for $\varphi = 0°$ and $180°$, there were taken values starting from $\varphi = 10°$. For $\varphi = 180°$, there is a discontinuity, but the computer program for the graphical representation has linked the two limit points, obtaining a slopped line in the middle.

From Table 4.2, it appears that in the fast-growing zone to infinity, those subdomains are omitted as beyond the limit display.

Figures 4.56 and 4.57 show the successive positions of the mechanism, with different iteration steps, presenting parabola generation as envelope of a line.

It is clear that the variation of the radii lengths and the tangents to the points on the line *TT* are parallel to *xx*. It is also found that as *AB* approaches the *x*-axis, point *B* moves more rapidly and the generating tangent CD is closing on the vertical.

By decreasing Y_B, the parabola drawn becomes narrower, as in Fig. 4.58 ($Y_B =$ 40 mm), with smaller distances between branches.

If all lengths are considered equal, we obtain the parabola in Fig. 4.59, having symmetrical successive positions.

If Y_B is negative ($Y_B = -40$ mm), the same parabola is achieved for the same Y_B positive, but positioned under the axis *xx* (Fig. 4.60).

Table 4.2 Values for the coordinates of points B, C, and D

F_i	X_B	Y_B	X_C	Y_C	X_D	Y_D
30	138.5642	80	163.564	36.69862	118.5643	114.6411
40	95.34039	80	127.4796	41.69764	69.62893	110.6418
50	67.12806	80	105.4302	47.86045	36.48633	105.7116
60	46.18811	80	89.48926	54.9998	11.54713	100.0001
70	29.11771	80	76.10225	62.89876	−8.469972	93.68088
80	14.10624	80	63.34659	71.31735	−25.28606	86.94601
90	9.187E−05	80	50.0001	79.99975	−39.99991	80.0001
100	−14.10605	80	35.13438	88.68216	−53.49839	73.05418
110	−29.11749	80	17.86723	97.10077	−66.70523	66.31929
120	−46.18785	80	−2.886448	104.9998	−80.82892	60.00009
130	−67.12775	80	−28.82536	112.1392	−97.76959	54.28858
140	−95.33991	80	−63.20031	118.302	−121.0515	49.35831
150	−138.563	80	−113.5632	123.3011	−158.5636	45.35905
210	138.565	80	113.5653	123.3014	158.5648	45.35891
220	95.34085	80	63.20172	118.3024	121.0522	49.35812
230	67.1284	80	28.82642	112.1397	97.77006	54.28836
240	46.18836	80	2.887249	105.0003	80.82929	59.99985
250	29.11792	80	−17.86659	97.10136	66.70556	66.31903
260	14.10643	80	−35.1339	88.68274	53.49871	73.0539
270	3.042E−04	80	−49.9997	80.00038	40.00031	79.9998
280	−14.10584	80	−63.3463	71.31797	25.28651	86.94572
290	−29.11727	80	−76.10203	62.89936	8.470516	93.68061
300	−46.18761	80	−89.48907	55.00033	−11.54649	99.99982
310	−67.12745	80	−105.4299	47.86092	−36.48553	105.7114
320	−95.33952	80	−127.4792	41.69803	−69.62786	110.6417
330	−138.5626	80	−163.563	36.69893	−118.5624	114.6409

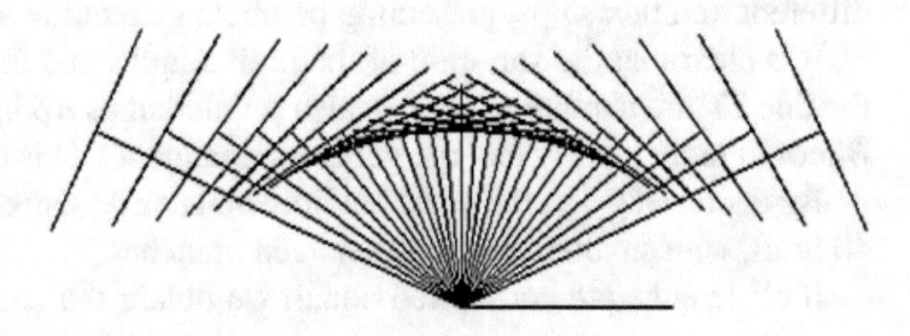

Fig. 4.56 Successive positions of the mechanism for $Y_B = 80$, $BD = 40$, $BC = 50$ mm

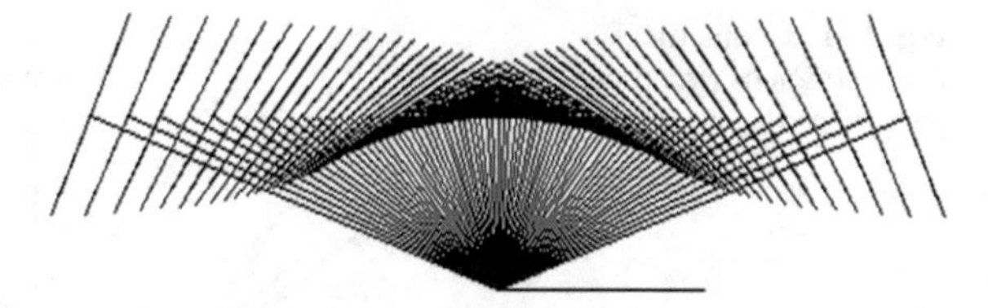

Fig. 4.57 Successive positions for $Y_B = 80$, $BD = 40$, $BC = 50$ mm, with a small step

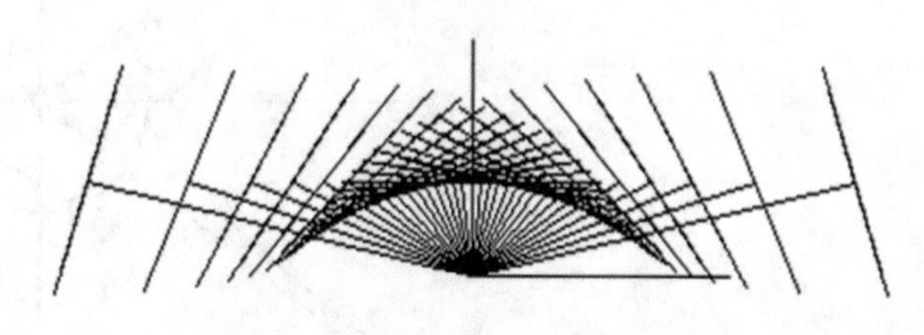

Fig. 4.58 Successive positions for $Y_B = 40$ mm

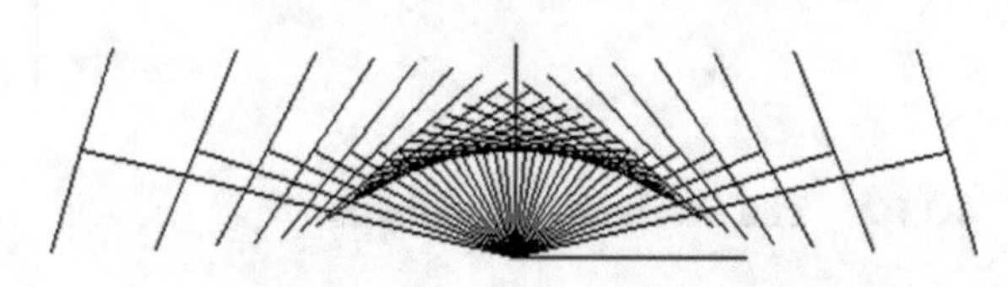

Fig. 4.59 Successive positions for $Y_B = BD = BC = 50$ mm

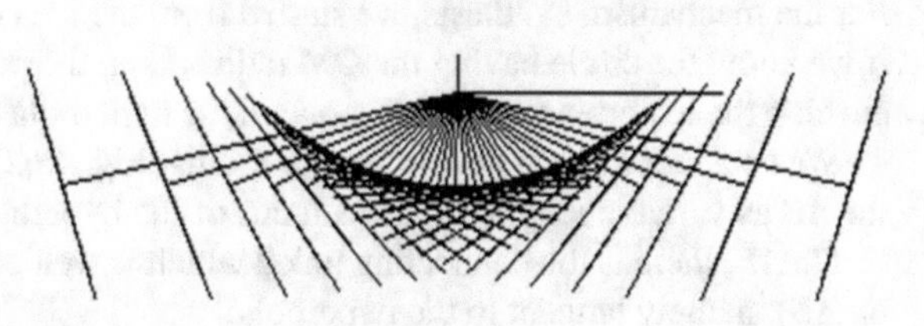

Fig. 4.60 Successive positions for $Y_B = -40$ mm

4.3 Mechanisms for Generating Hyperbolas

4.3.1 Case of Hyperbola Based on Pedal Curve

4.3.1.1 Geometrical Considerations

We started from Fig. 4.61 [24] where it is shown that the pedal curve of a hyperbola having the focuses F and F′ is the locus of the M point, resulted by tracing a perpendicular line from F on the tangent through P to the hyperbola. The wanted pedal curve is the Γ circle.

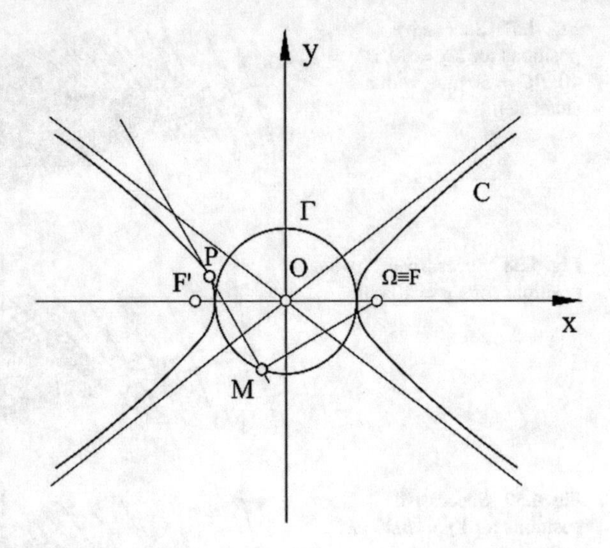

Fig. 4.61 Geometrical construction of hyperbola's pedal curve

4.3.1.2 The Mechanism Synthesis

For the mechanism synthesis, we started from the idea of the inverted problem, i.e., if we know the circle having the OM radius, then the perpendicular line on FM will envelop the hyperbola, so we may obtain the hyperbola as an envelope of a line [10].

We consider a circle having the *CB* radius (Fig. 4.62) and the straight line having the *AB* as variable length, as *A* is a focus of the hyperbola [15].

On *AB*, there is the connecting link 2 which is welded to *EBF* line (perpendicular on *AB*), namely tangent to the hyperbola.

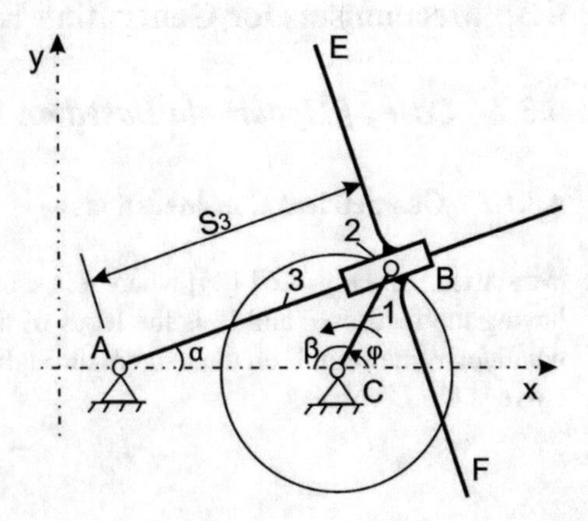

Fig. 4.62 Mechanism obtained

4.3.1.3 The Mechanism Analysis

Structurally, the mechanism type is R-RPR [14]

Based on Fig. 4.62, we write Eqs. (4.30–4.33):

$$x_B = x_C + CB\cos\varphi = x_A + S_3\cos\alpha \tag{4.30}$$

$$y_B = CB\sin\varphi = y_A + S_3\sin\alpha \tag{4.31}$$

$$S_3 = \sqrt{AC^2 + CB^2 - 2AC\cdot CB\cos\alpha} \tag{4.32}$$

$$\tan\alpha = \frac{y_B - y_A}{x_B - x_A} \tag{4.33}$$

4.3.1.4 Results Obtained

Figure 4.63 shows the computer-generated mechanism for a given position. The length of the elements *EB* and *BF* was considered big enough ($EB = BF = 200$ mm) for enveloping the two branches of the hyperbola, imposing the appropriate length of the tangent.

We selected: $X_A = 20$ mm; $AC = 50$ mm; $CB = 30$ mm.

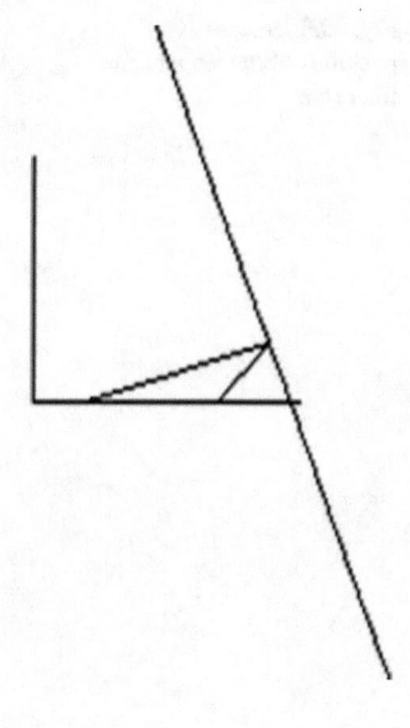

Fig. 4.63 Mechanism for a position

In Fig. 4.64, it is shown the generated hyperbola (only the successive positions of the tangent were drawn). It is noted that the left branch has fewer tangents than the right one.

In Fig. 4.65, it is shown the hyperbola resulted in the same data, but with a finer step, as this time both of the branches are well defined.

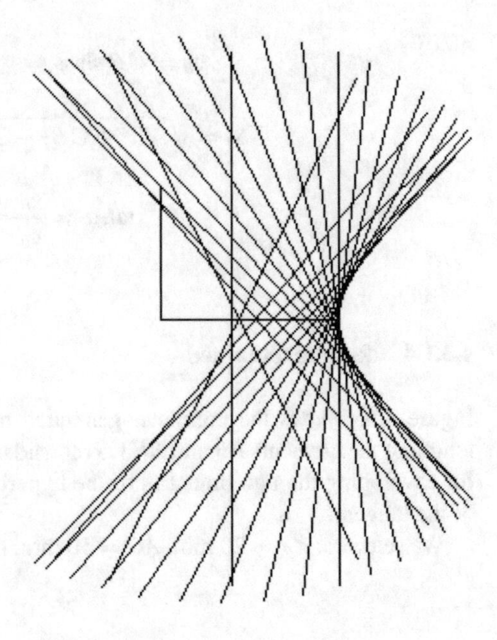

Fig. 4.64 Successive positions of the tangent

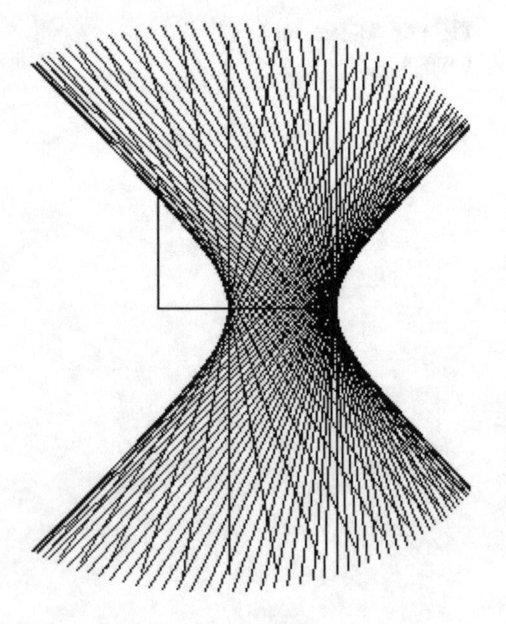

Fig. 4.65 Successive positions of the tangent for finer step

In Fig. 4.66, the curves for X_B and Y_B are continuous, so there are no blockages, and therefore, the mechanism works the entire cycle from $\varphi = 0$ to $\varphi = 360°$.

The S_3 displacement (Fig. 4.67) has a symmetrical variation, except for the starting area from $\varphi = 0$. This is also a continuous curve, so the movement is continuous.

The variations of the coordinates of E and F points (Fig. 4.68) have symmetries and similarities for the ordinates, but different curves for the abscissas.

In Fig. 4.69, there is E's trajectory, and in Fig. 4.70, there is F's trajectory. The trajectories are similar, but delayed with respect to the unique system of axes.

In Fig. 4.71, there is the mechanism having the same constructive data, but with a greater cycled step, because we may see the length of all the elements, finding

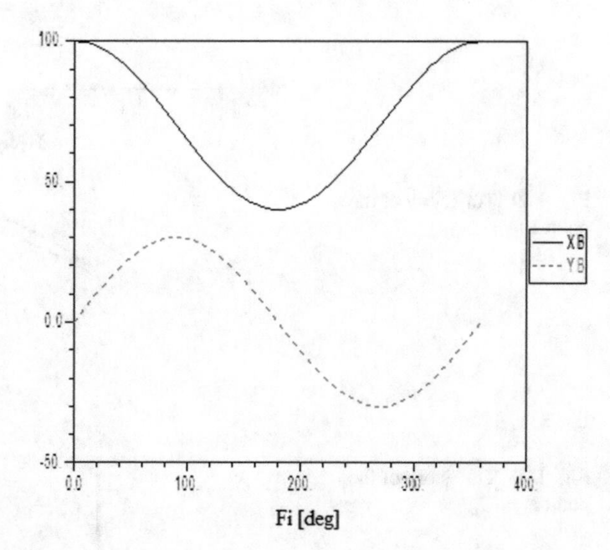

Fig. 4.66 Curves for X_B and Y_B

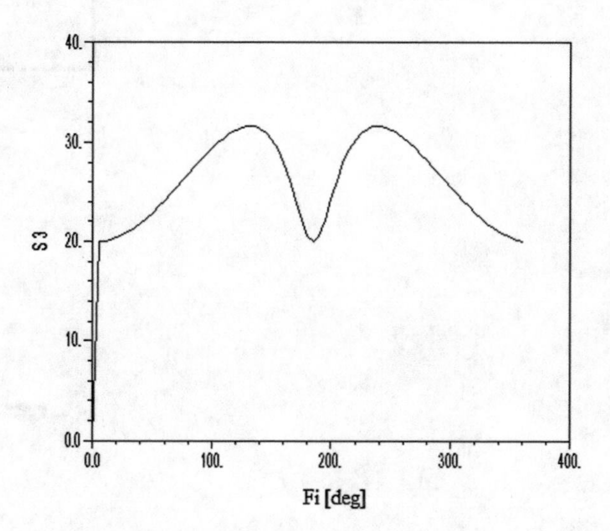

Fig. 4.67 S_3 (φ) displacement

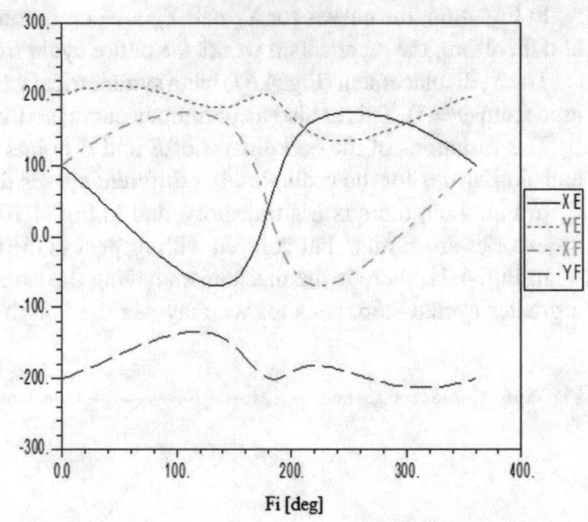

Fig. 4.68 Variations of E and F points' coordinates

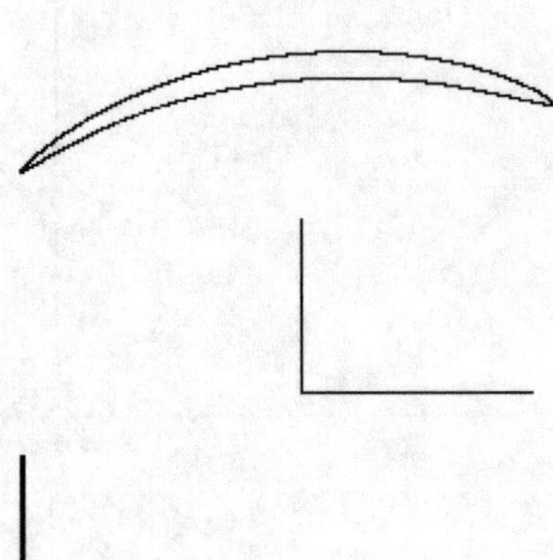

Fig. 4.69 Trajectory of the point E

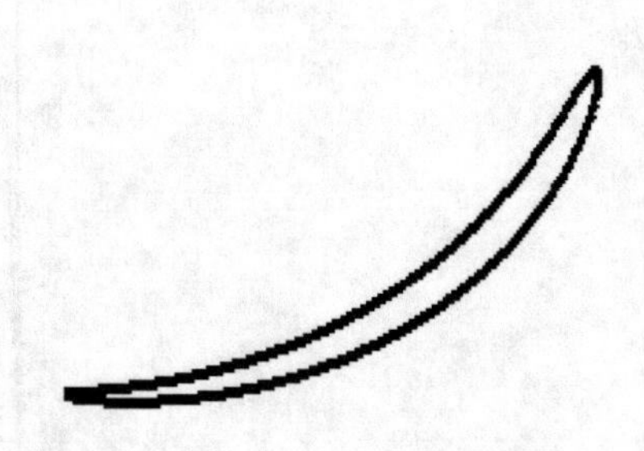

Fig. 4.70 Trajectory of the point F

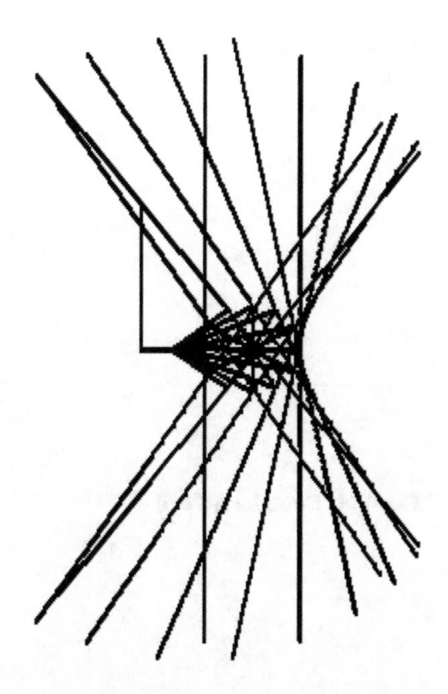

Fig. 4.71 Mechanism's successive positions

that the *CB* element executes a complete rotation, and the *AB* element has only an oscillation movement.

Next, we increased the distance *AC*, from 50 mm to 80 mm (Fig. 4.72) and made the cycle with a finer step, resulting in a narrower hyperbola.

Changing *AC* again, decreasing at the value $AC = 10$ mm, we obtained an ellipse (Fig. 4.73), as the focus point is too close to the circle center, C.

For $AC = 0$, we obtained the circle in Fig. 4.74 because the focus point fits the circle center.

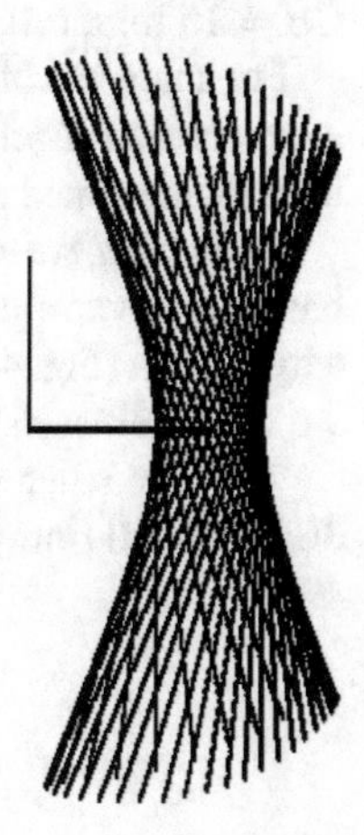

Fig. 4.72 Narrow hyperbola

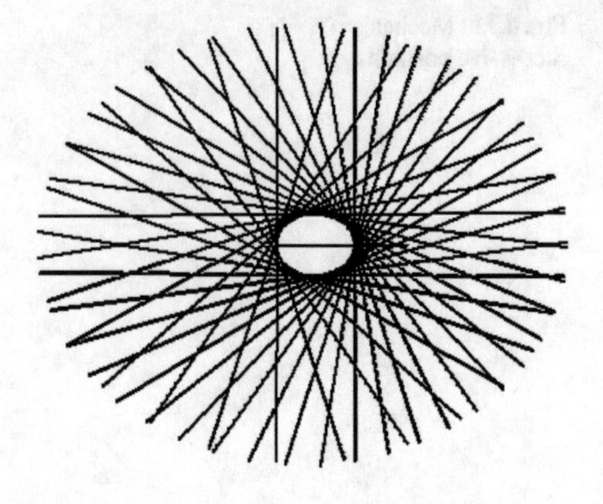

Fig. 4.73 Ellipse for $AC = 10$ mm

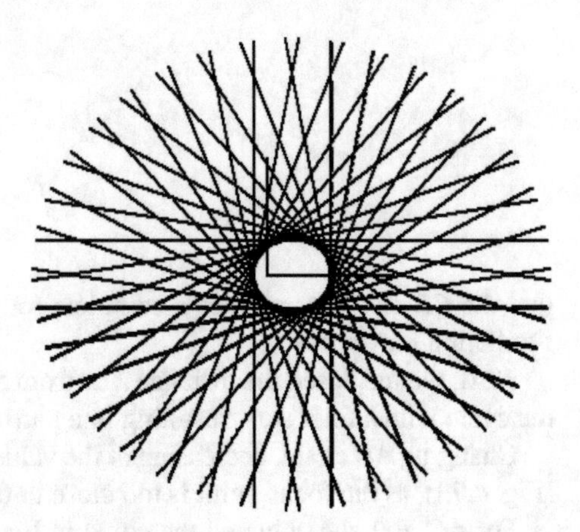

Fig. 4.74 Circle for $AC = 0$

Next, we changed the data to the following values: $X_A = 20$ mm; $AC = 50$ mm; CB = 15 mm, resulting in a narrow hyperbola in Fig. 4.75.

For a new set of data, i.e., $X_A = 20$; $AC = 50$; $CB = 35$; $Y_A = 20$ [mm], it results a hyperbola rotated clockwise (Fig. 4.76), and for $Y_A = -20$ mm, we obtained the hyperbola rotated counterclockwise (Fig. 4.77).

Further on, we considered the point A to be to the left of the system of axes, i.e., X_A has a negative value $X_A = -20$ mm ($AC = 50$; $CB = 30$; $Y_A = 0$ [mm]), resulting in a hyperbola (Fig. 4.78) and similar to the normal ones (Figs. 4.72 and 4.75), because we only increased the distance from C to A.

The case when point C is positioned to the left of point A, for $X_A = 20$; $AC = -30$; CB = 30 [mm], there is no more hyperbola (Fig. 4.79).

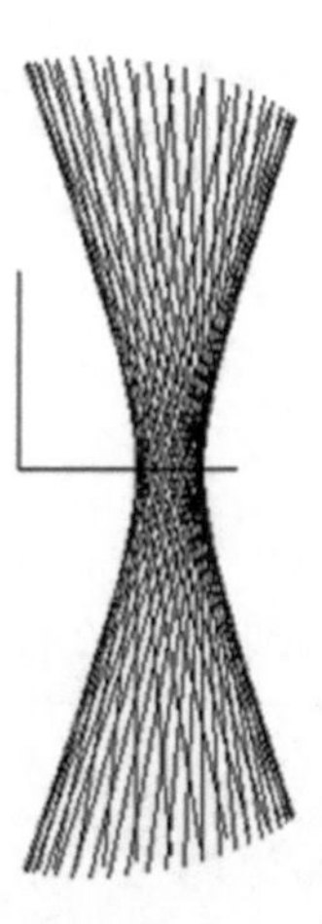

Fig. 4.75 Successive positions for $X_A = 20$ mm; $AC = 50$ mm; $CB = 15$ mm

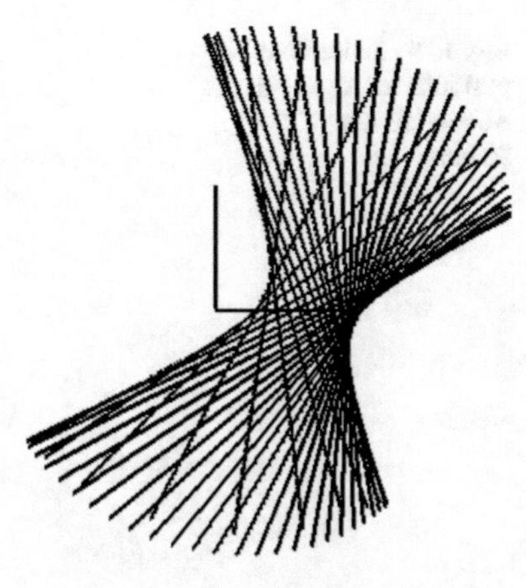

Fig. 4.76 Rotated clockwise hyperbola

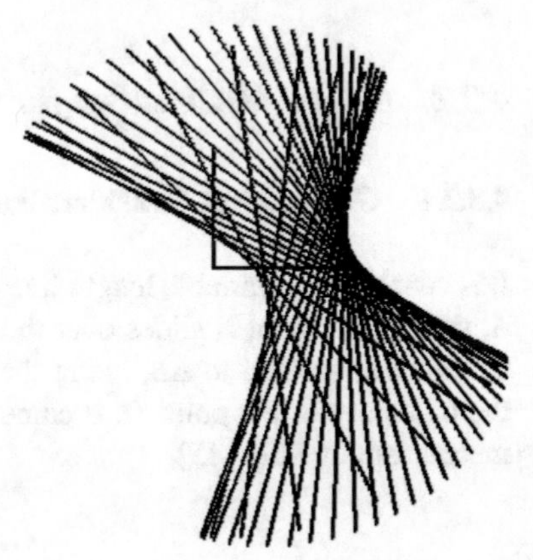

Fig. 4.77 Rotated counterclockwise hyperbola

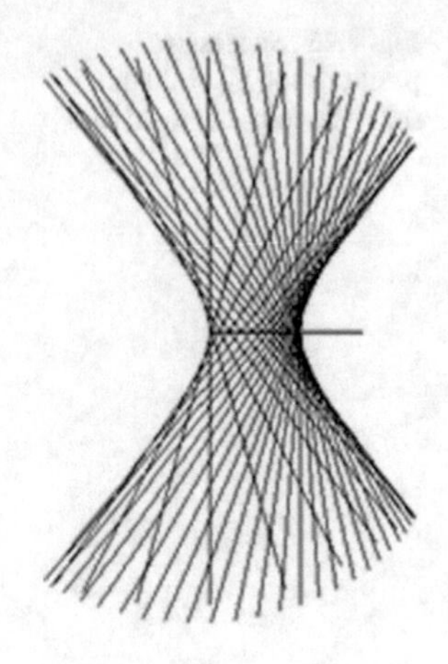

Fig. 4.78 Successive positions ($X_A = -20$; $AC = 50$; $CB = 30$; $Y_A = 0$)

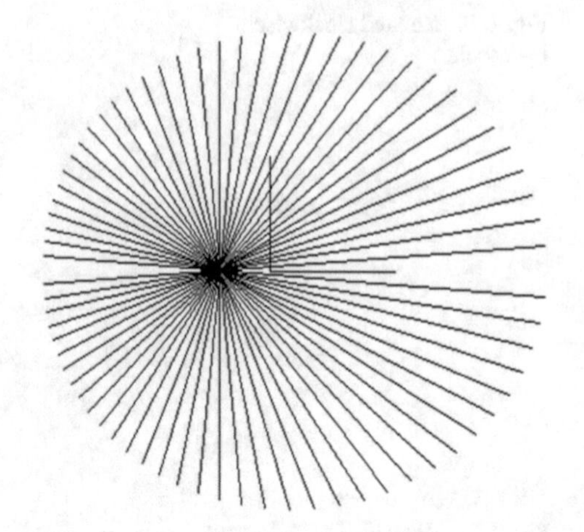

Fig. 4.79 Successive positions for $X_A = 20$ mm; $AC = -30$ mm; $CB = 30$ mm

4.3.2 A New Mechanism for Generating Hyperbola

4.3.2.1 Geometrical Considerations

It is considered a variable length line *AB* (Fig. 4.80), which rotates around the point *A*, so that the point *B* slides over the fixed directrix *EB*. The fixed length segment *CD* moves, parallel to *EB*, along the parallel to *AE* traced by *B*. It is looking for the trajectory of the point *D*. It came to this design looking for conchoidal tracing mechanism of Kulp [25].

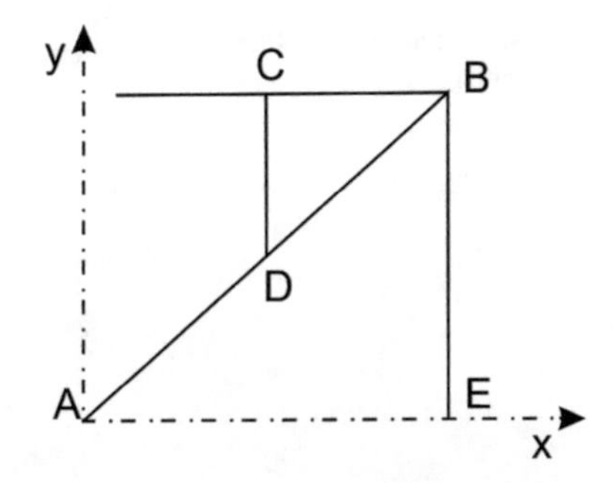

Fig. 4.80 Principle of generation

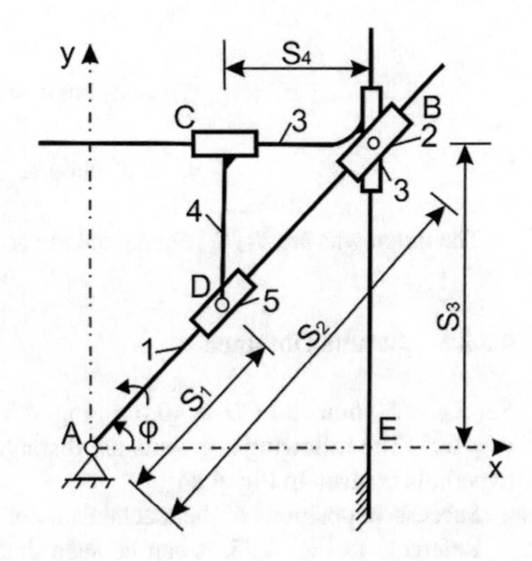

Fig. 4.81 Obtained mechanism

4.3.2.2 The Mechanism Synthesis

The mechanism that fulfills the described geometrical conditions has to allow the movement of *B* on the fixed vertical *EB* (Fig. 4.81), but also the movement of *AB*. This problem was solved by placing two sliders with rotational motion relative to each other, at point *B* [9]. The slider 3 is welded to *BC* bar, so that it is permanently parallel to the abscissa.

4.3.2.3 The Mechanism Analysis

The mechanism has five elements and seven fifth class pairs, so its degree of mobility is 1.

The structural scheme and decomposition into kinematic groups (Fig. 4.82) show that the mechanism is of the type R-PRP-PRP [8].

On the basis of Fig. 4.82, we write Eqs. (4.34)–(4.37):

$$S_2 \cos \varphi = \text{const.} \tag{4.34}$$

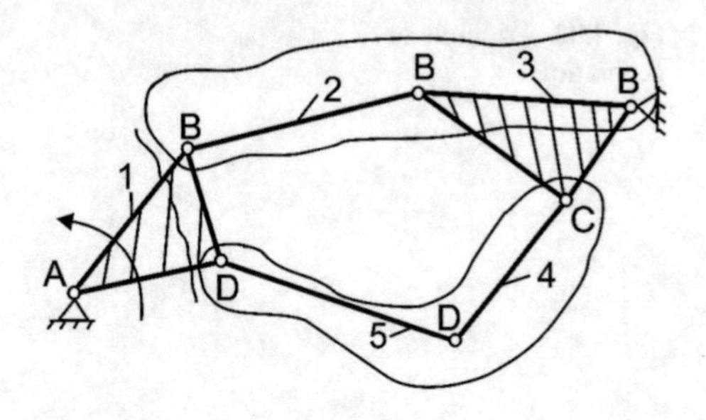

Fig. 4.82 Structure of the mechanism

$$y_B = S_2 \sin \varphi = S_3 = y_C \tag{4.35}$$

$$x_D = S_1 \cos \varphi = x_B - S_4 \tag{4.36}$$

$$y_D = S_1 \sin \varphi = y_B - CD \tag{4.37}$$

The unknowns are S_1, S_2, S_3, S_4 and the coordinates of points C and D.

4.3.2.4 Results Obtained

For $X_B = 50$ mm and $CD = 40$ mm, Fig. 4.83 shows the resulting mechanism for $\varphi = 50°$. The following segments are distinguished: *EB*, *BC*, *CD*, *AB*. The drawn hyperbola is given in Fig. 4.84.

Successive positions of the mechanism are shown in Fig. 4.85.

Referring to Fig. 4.85, it can be seen that the element AB (Fig. 4.81) rotates only in the trigonometric quadrants I and IV, which is explained by the fact that in the proximity of the *y*-axis, the point *B* tends to be infinite as the line EB is fixed. The program was set to skip instructions for the cases when Y_B increases more (in absolute values).

In Fig. 4.86, there are shown the variations of the coordinates of the generating point D, for a whole functioning cycle of the mechanism.

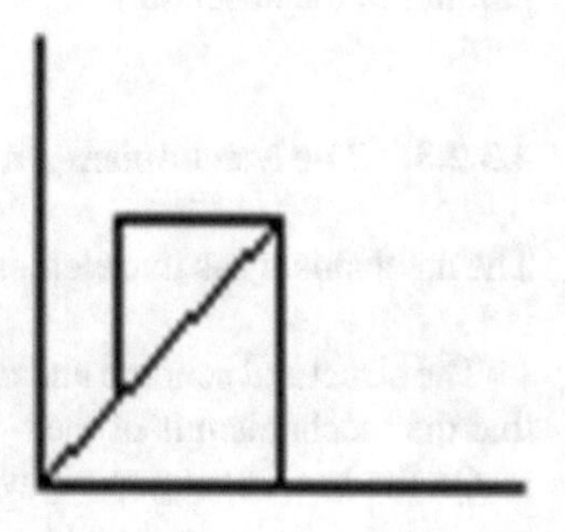
Fig. 4.83 Position of the mechanism

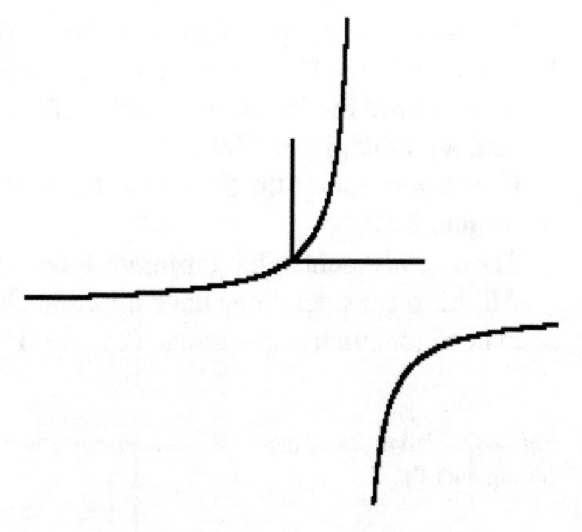

Fig. 4.84 Obtained hyperbola

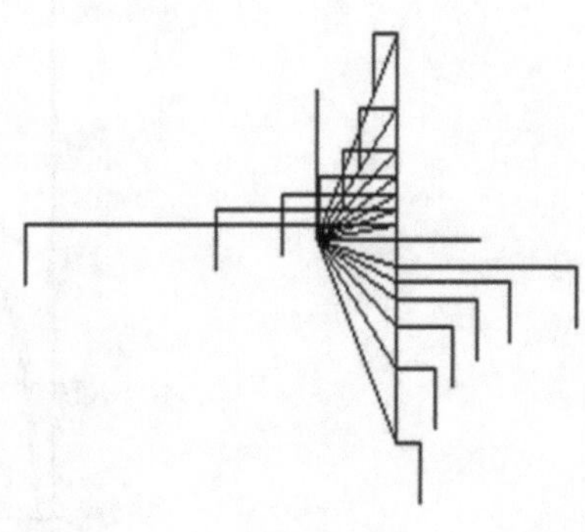

Fig. 4.85 Successive positions of the mechanism

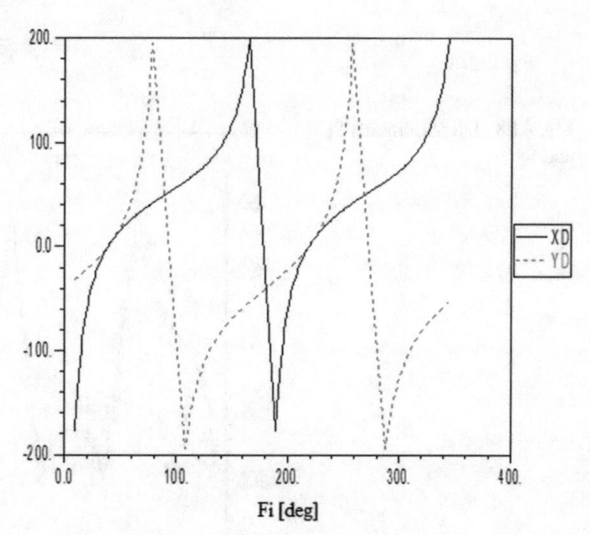

Fig. 4.86 Variations of the coordinates of point *D*

It is noted the jumps around $\varphi = 90°$ and 270° for Y_D and around 180° for X_D. Referring to Fig. 4.87, it is noted that the coordinates of point C (with $X_C = X_D$ and $Y_C = Y_D$) have the jumps too, but in different areas: Y_B at φ around 90° and 270°, X_C and X_D around $\varphi = 180°$.

Figure 4.88 shows the S_1 jumps at $\varphi = 90°$, 180°, 270°, and only S_2 jumps for $\varphi = 90°$ and 270°.

The displacement S_3 has the jumps for $\varphi = 90, 270$, and S4 for $\varphi = 180°$ (Fig. 4.89).

All these considerations have imposed checking the functionality of the mechanism in trigonometric quadrants. For $\varphi = 0° \ldots 90°$ (Fig. 4.90), the upper branch of

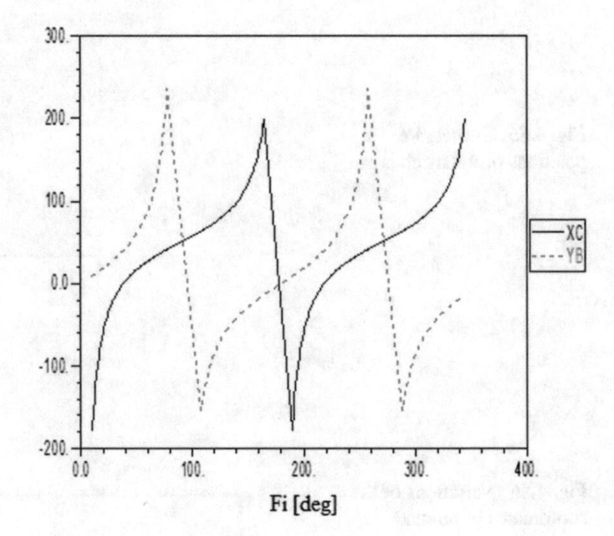

Fig. 4.87 Variations of the coordinates Y_B, X_C

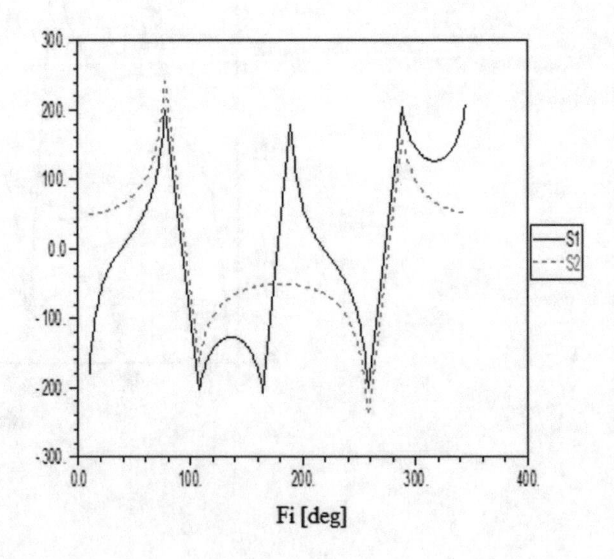

Fig. 4.88 Displacements S_1 and S_2

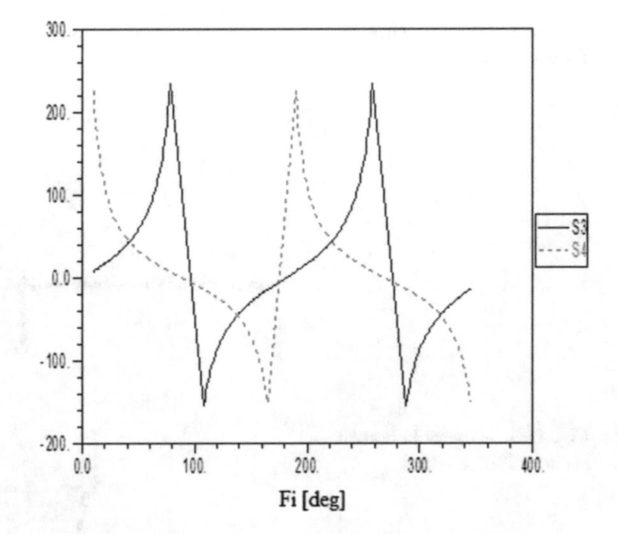

Fig. 4.89 Displacements S_3 and S_4

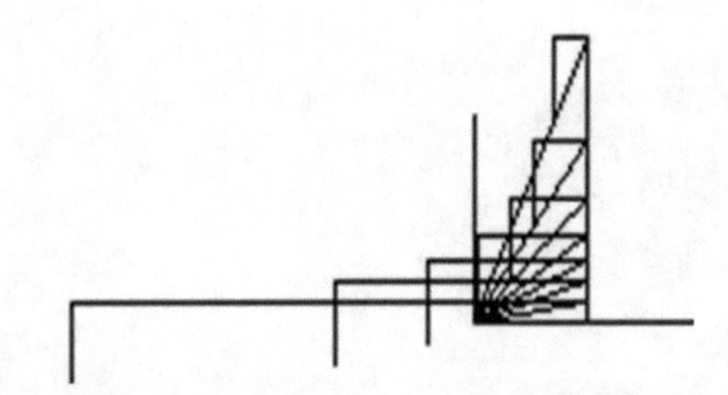

Fig. 4.90 Upper branch of the hyperbola for $\varphi = 0° \ldots 90°$

the hyperbola is generated. As it is shown also in Fig. 4.86, for small values of φ, X_D is negative, then the point D is on the left side of the ordinate. The slope of the curve X_D of Fig. 4.86 is great in this subinterval, which makes it as in Fig. 4.90. Successive positions of the mechanism appear less, although the cycled step was the same. The hyperbola is therefore drawn from quadrant II to quadrant I.

The increase of φ between 90° and 180° (Fig. 4.91) Y_D is negative (see Fig. 4.86), so that the hyperbola lower branch is drawn, starting from the lowest D point.

For $\varphi = 180° \ldots 270°$, Fig. 4.92 results as identical to Fig. 4.90, which means this runs the same branch of the hyperbola for the second time.

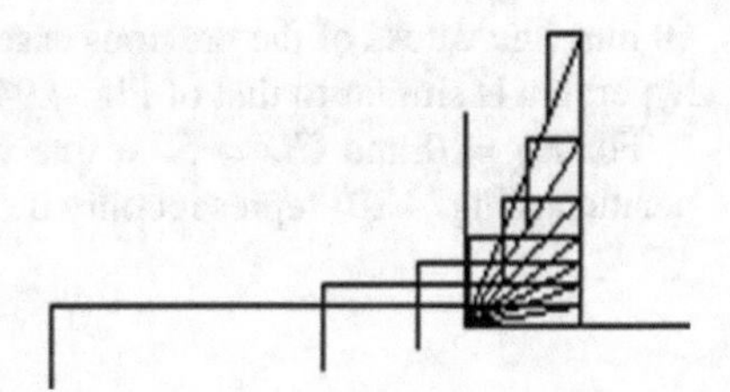

Fig. 4.91 Hyperbola branch for $\varphi = 90° \ldots 180°$

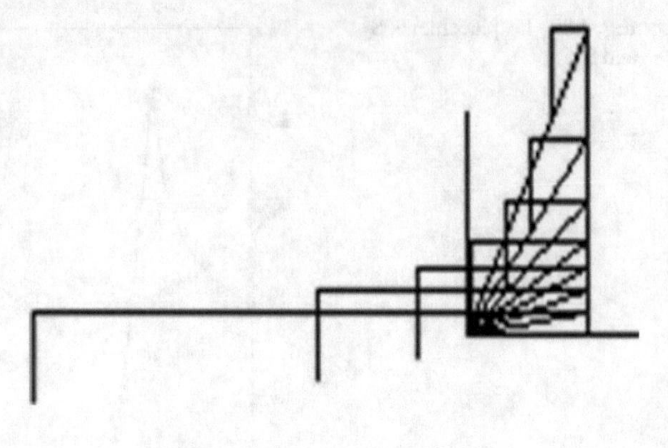

Fig. 4.92 Hyperbola branch for $\varphi = 180° \ldots 270°$

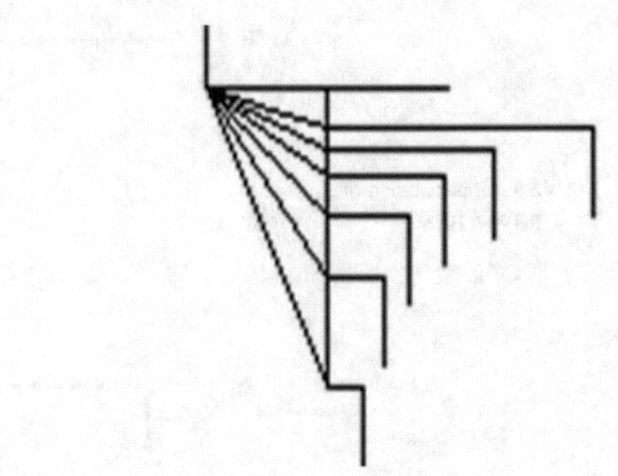

Fig. 4.93 Hyperbola branch for $\varphi = 270° \ldots 360°$

Similarly, for $\varphi = 270° \ldots 360°$ Fig. 4.93 is obtained as identical to Fig. 4.91, which means the other branch is run through for the second time.

It results from the foregoing that the mechanism works only in the areas where successive positions are traced for the element AB because the line EB, on which B moves, is fixed. However, the mechanism generates the two branches of the hyperbola.

Next, the influence of the mechanism input data on the shape and size of the generated hyperbola was studied.

Thus, for $X_B = 60$ mm and $CD = 30$ mm it was generated the hyperbola in Fig. 4.94.

In Figs. 4.95 and 4.96, the curve and the mechanism for $X_B = 30$ mm, $CD = 60$ mm (the values of the previous case were inverted to each other) are shown. The hyperbola is similar to that of Fig. 4.94, but has branches wider to the outside area.

For $X_B = 0$ and $CD = 50$ a line was obtained (Fig. 4.19), and the successive positions (Fig. 4.20) represent only the CD segment.

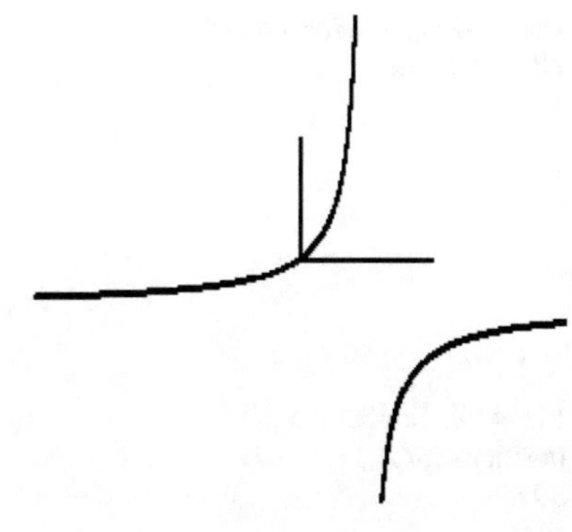

Fig. 4.94 Hyperbola for X_B = 60 mm, CD = 30 mm

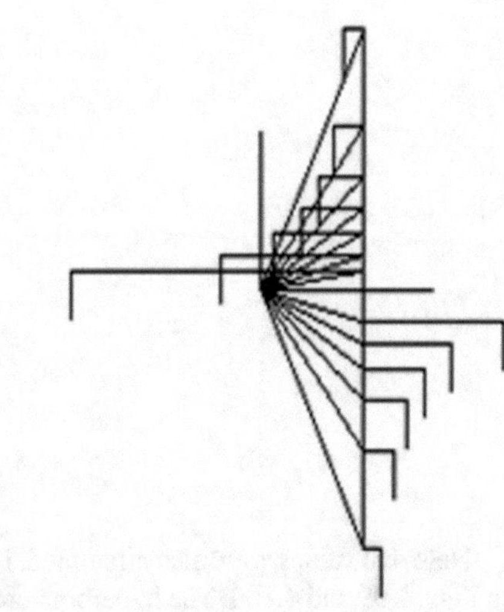

Fig. 4.95 Curve for X_B = 30 mm, CD = 60 mm

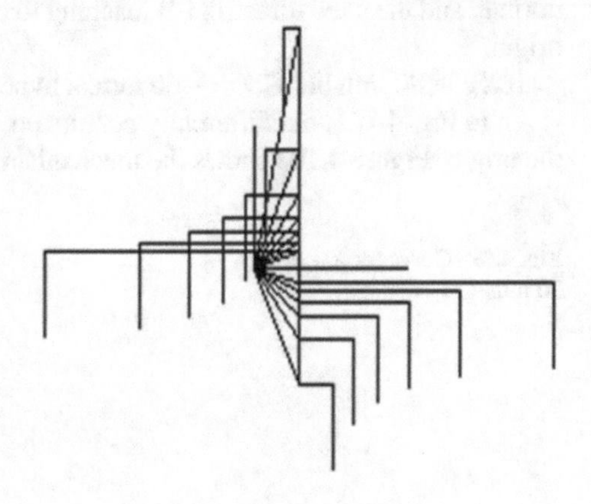

Fig. 4.96 Successive positions of the mechanism for X_B = 30 mm, CD = 60 mm

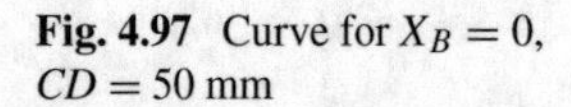

Fig. 4.97 Curve for $X_B = 0$, $CD = 50$ mm

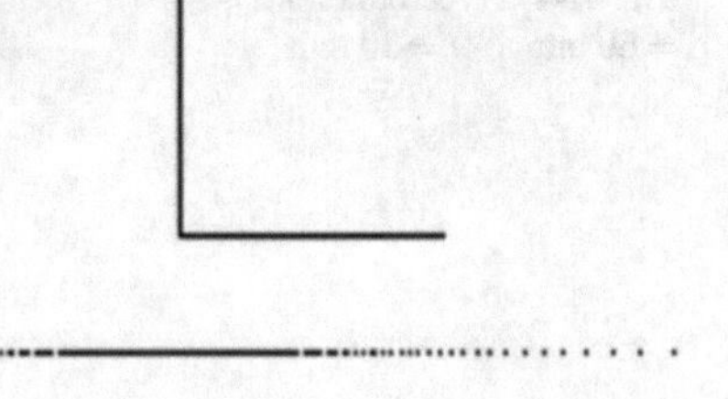

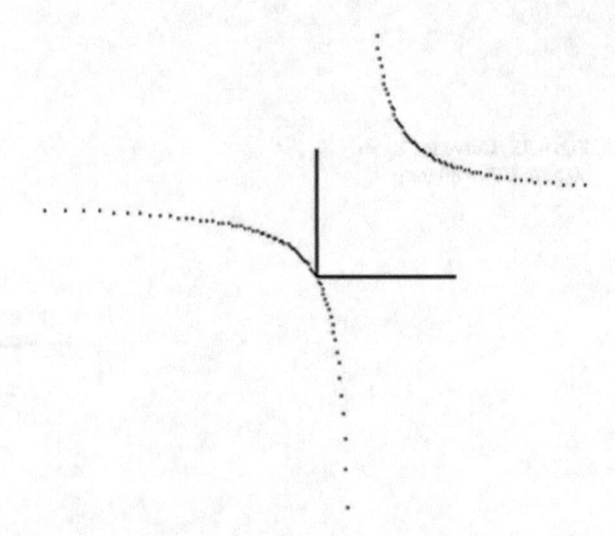

Fig. 4.98 Successive positions for $X_B = 0$, $CD =$ 50 mm

Negative values were also attempted, i.e., $X_B = -30$ mm, $CD = 60$ mm, resulting in Figs. 4.99 and 4.100. The hyperbola changed its position and orientation, which is normal, and the fixed directrix EB reaching to the left side of the coordinate system's origin.

If $X_B = 30$ mm but $CD = -60$ mm, a hyperbola is obtained with the orientation given in Fig. 4.101, but differently positioned, the left branch crossing also through the origin. Figure 4.102 shows the mechanism for different positions.

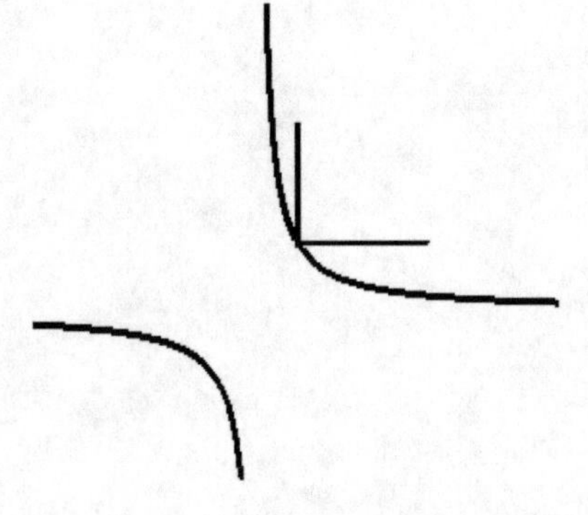

Fig. 4.99 Curve for $X_B = -$ 30 mm, $CD = 60$ mm

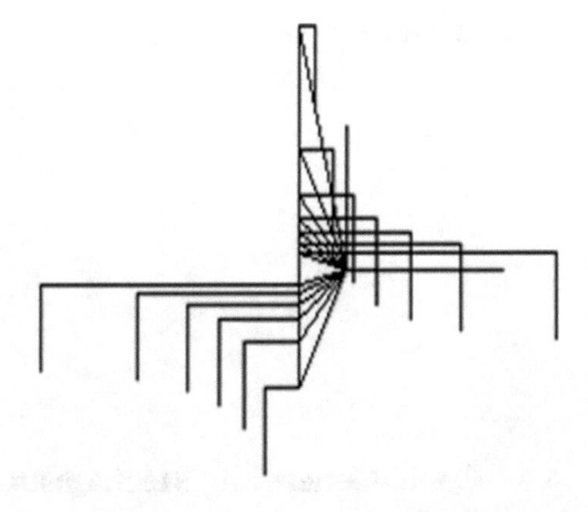

Fig. 4.100 Successive positions for $X_B = -30$ mm, $CD = 60$ mm

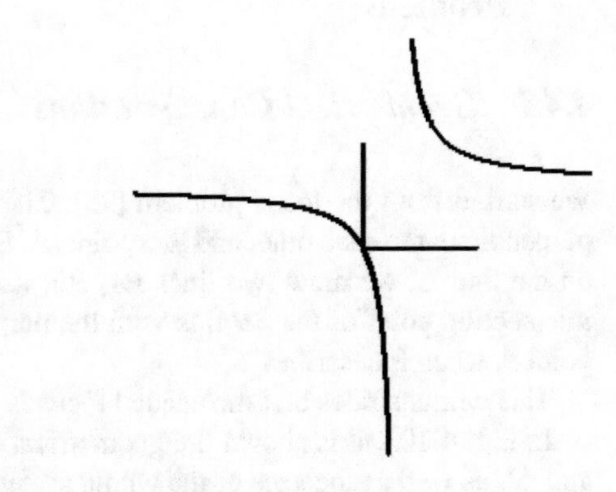

Fig. 4.101 Curve for $X_B = 30$ mm, $CD = -60$ mm

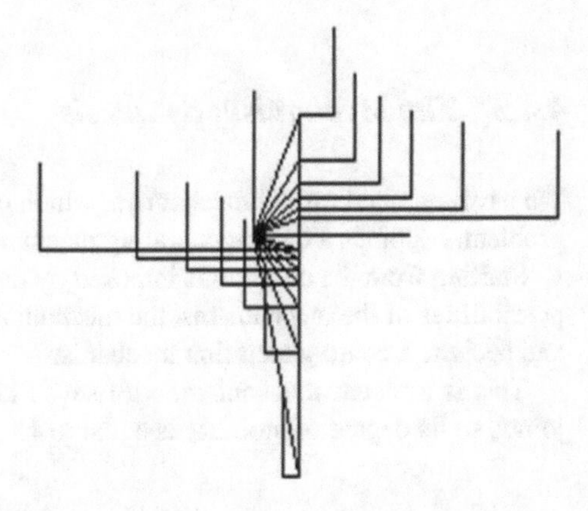

Fig. 4.102 Successive positions for $X_B = 30$ mm, $CD = -60$ mm

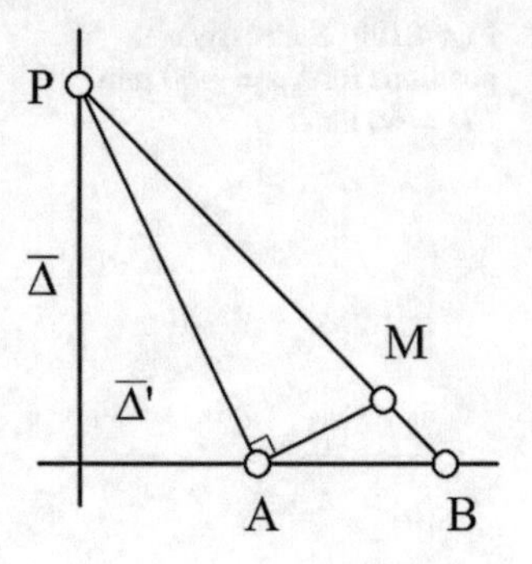

Fig. 4.103 Locus problem

4.4 Conic-Generating Mechanism Based on a Locus Problem

4.4.1 Geometrical Considerations

We started from the locus problem [25]: There are given two straight lines Δ, Δ', perpendicular to each other and two points A, B located on Δ'; from a point P situated on the line Δ, we draw two lines *PA*, *PB*; it is required to find the position of the intersection point of the *PB* line with the perpendicular to the line *PA*, drawn from point A when P describes Δ'.

This demonstrates that the needed locus is a conic.

In Fig. 4.103, it is shown the geometrical construction, considering the lines Δ and Δ', as well as the axes of the whole system [10].

4.4.2 The Mechanism Synthesis

We have searched for the mechanism, which constructed, based on this geometrical problem, becomes a conic-generating mechanism [6, 13, 16, 21].

Starting from the conditions imposed by the problem and taking into account the possibilities of the mechanisms, the mechanism shown in Fig. 4.104 was conceived and became a conic-generating mechanism.

This is a planar mechanism, with seven kinematic elements and ten fifth class joints, so its degree of mobility is equal to 1.

4.4.3 The Mechanism Analysis

Using the contours method, we write Eqs. (4.38, 4.39) for the contour PB to determine the angle α and the distance S_7:

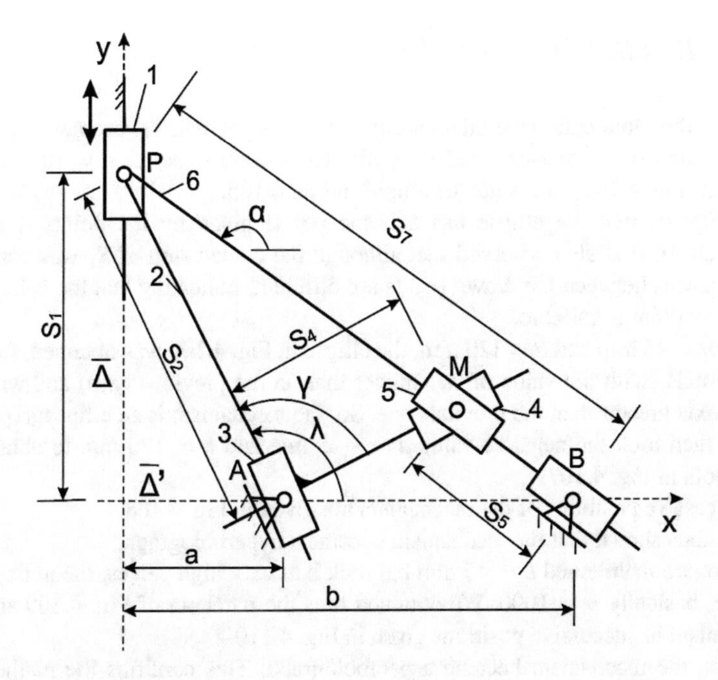

Fig. 4.104 Constructed mechanism

$$x_P = x_B + S_7 \cos\alpha = 0 \tag{4.38}$$

$$y_p = S_1 = S_7 \sin\alpha \tag{4.39}$$

Further on, we write Eqs. (4.40, 4.41) for the contour *PA* to determine the angle λ and the distance S_2:

$$x_P = x_A + S_2 \cos\lambda = 0 \tag{4.40}$$

$$y_p = S_1 = S_2 \sin\lambda \tag{4.41}$$

Next, we determine S_4, S_5, x_M, y_M from Eqs. (4.42, 4.43), written for the contour AMB:

$$x_M = x_A + S_4 \cdot \cos(\lambda - \gamma) = x_B + S_5 \cdot \cos\alpha \tag{4.42}$$

$$y_M = y_A + S_4 \cdot \sin(\lambda - \gamma) = y_B + S_5 \cdot \sin\alpha \tag{4.43}$$

4.4.4 Results Obtained

Initially, the input data were taken according to the geometrical problem, meaning points A and B on the x-axis and $\gamma = 90°$. There were selected: $a = 10$ mm, $b =$ 100 mm (Fig. 4.104). It results the ellipse in Fig. 4.105.

It appears that the ellipse has extreme points given by the points A and B (Fig. 4.103). It is also observed that although the cycled step of S_1 was constant, the distances between the drawn points are different, indicating that the velocity of the tracer point is variable.

For $a = 45$ mm and $b = 120$ mm, the ellipse in Fig. 4.106 was obtained, furthest from y-axis (with the value of "a" higher than in the previous case) and with the minor axis greater than the previous one. So, the mechanism is an ellipsograph.

We then took the negative value $a = -45$ mm and $b = 120$ mm to obtain the hyperbola in Fig. 4.107.

Successive positions of the mechanism are given in Fig. 4.108.

It is thus shown that the mechanism became a hyperbolograph.

Next, we maintained $a = 45$ mm but took b at very high values, theoretically to infinite, basically $b = 1000$. We obtained thus the parabola of Fig. 4.109 and the mechanism in successive positions given in Fig. 4.110.

Thus, the mechanism became a parabolograph. This confirms the mathematical considerations [25] and the accuracy of the designed mechanism, as being a conicograph (conic-generating) mechanism.

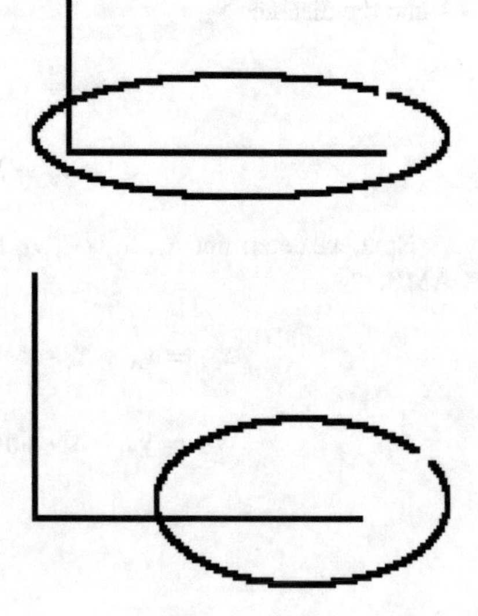

Fig. 4.105 Resulted ellipse

Fig. 4.106 New resulted ellipse

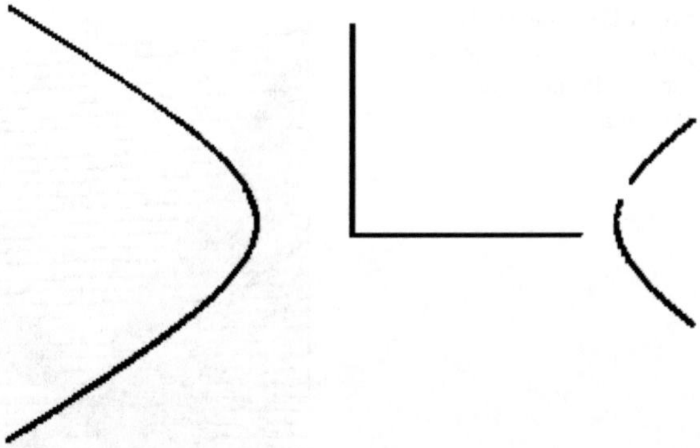

Fig. 4.107 Hyperbola for $a = -45$ mm, $b = 120$ mm

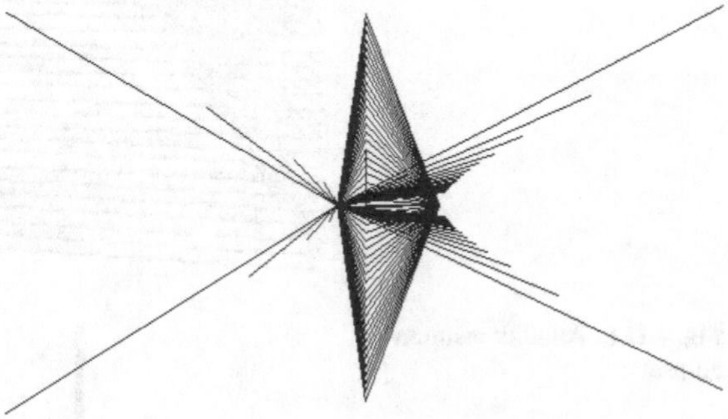

Fig. 4.108 Successive positions of the mechanism

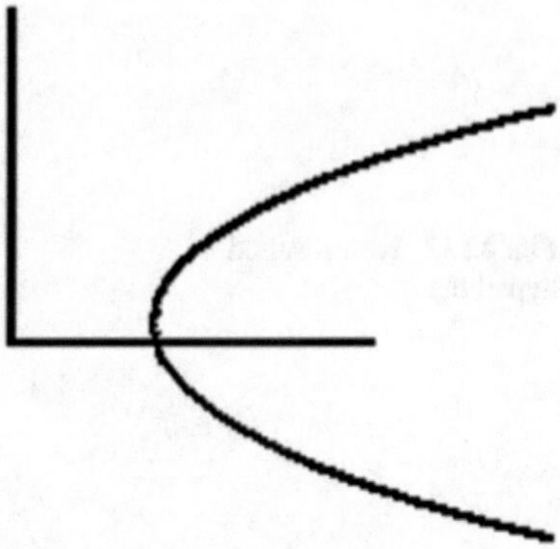

Fig. 4.109 Parabola for $a = 45$ mm, $b = 1000$ mm

With our developed computer program, other solutions were tested, exceeding the initial data of the problem. Thus, it was taken $\gamma = 45°$, $Y_A = 20$ mm, $Y_B = -10$ mm. For $a = 30$ mm and $b = 70$ mm, the ellipse in Fig. 4.111 resulted, which has specifically another position toward the cases in Figs. 4.105 and 4.106.

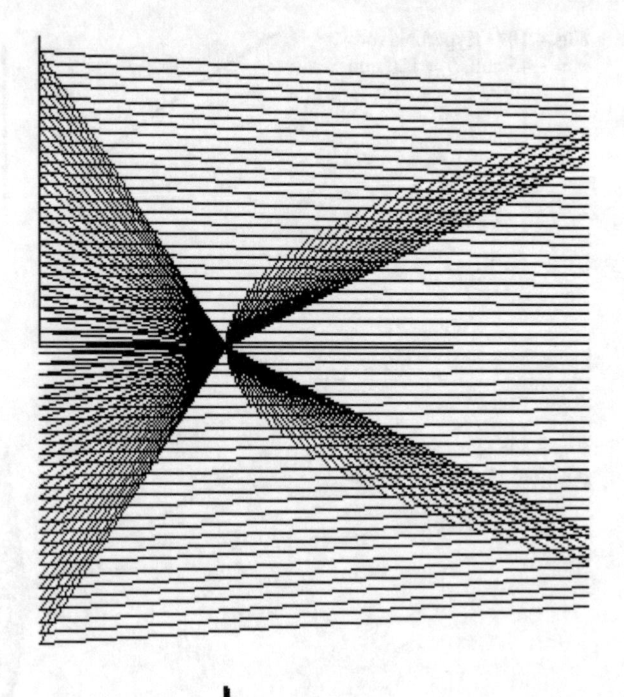

Fig. 4.110 Successive positions of the mechanism for $a = 45$ mm, $b = 1000$ mm

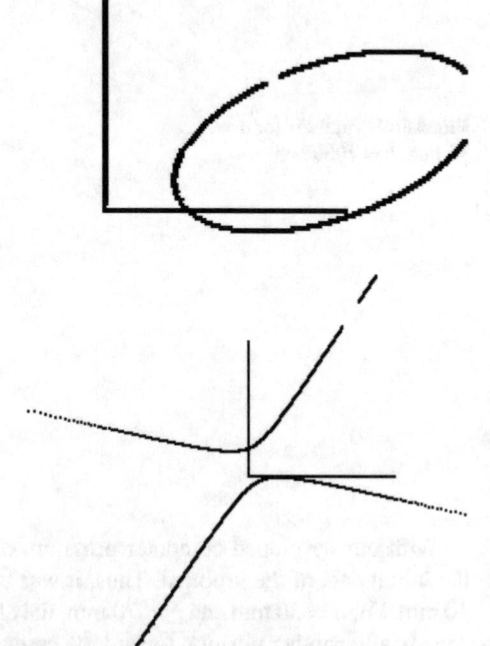

Fig. 4.111 Another resulted ellipse

Fig. 4.112 New resulted hyperbola

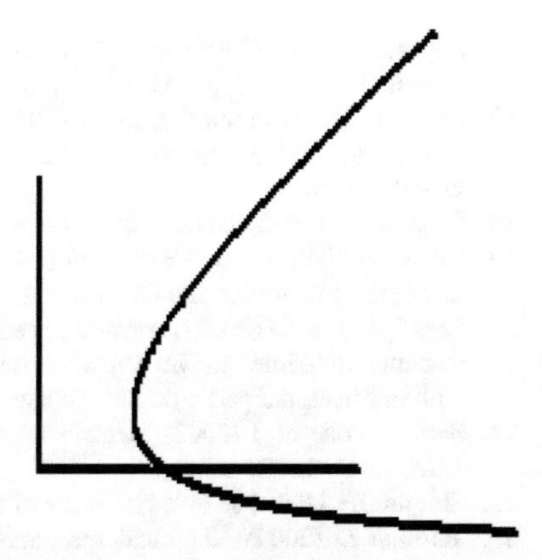

Fig. 4.113 New resulted parabola

For $a = -30$ mm, $b = 70$ mm, a hyperbola resulted, given in Fig. 4.112, more inclined than the normal one. For $a = 30$ mm and $b = 1000$ mm, the parabola from Fig. 4.113 was obtained, also more inclined than the normal one.

References

1. Artobolevskii II (1959) Teoria mehanizmov dlia vosproizvedenia ploskih crivâh. Izd. Academii Nauk SSSR Moskva
2. Artobolevskii II (1951) Mehanizmî, vol II. Izd. Academii Nauk, SSSR, Moskva
3. Bernhard A (1993) Geometrie proiectivă. Perceperea spaţiului prin desen, Ed. Tehnică, Bucuresti
4. Creangă I (1953) Curs de geometrie analitică. Editura Tehnică, Bucureşti
5. Luca L, Popescu I (2004) Asupra unui nou mecanism parabolograf. In: Mecanisme şi manipulatoare, nr. x/2004, 5pp
6. Luca L, Popescu I, Cârţână L (2005) The structure and kinematics of Kaminski's conicograph, In: Proceedings, Sections 7, Transcom 2005, University of Zilina, Slovak Republic, pp 129–131
7. Luca L, Popescu I (2004) Sinteza unui mecanism parabolograf. În: Construcţia de Maşini, nr. 1-2/2004, pp 117–118
8. Popescu I, Luca L, Cherciu M (2013) Structura şi cinematica mecanismelor. Aplicaţii, Editura Sitech, Craiova
9. Popescu I, Sass L (2001) Mecanisme generatoare de curbe, Ed. "Scrisul Românesc", Craiova
10. Popescu I (2013) Curbe şi suprafeţe estetice: geometrie, generare, aplicaţii. Editura Sitech, Craiova
11. Popescu I, Luca L, Mitsi S (2011) Geometria, structura şi cinematica unor mecanisme. Editura Sitech, Craiova
12. Popescu I, Luca L, Cherciu M (2011) Traiectorii şi legi de mişcare ale unor mecanisme. Editura Sitech, Craiova
13. Popescu I, Sass L, Luca L (2013) Nou mecanism conicograf. In: Construcţia de maşini, nr. 11/2013
14. Popescu I, Luca L (2004) Construcţia şi structura unui mecanism hiperbolograf. In: Proceedings scientific conference, 9th edn. Univesity of Târgu-Jiu, pp 39–44
15. Popescu I, Luca L (2012) Synthesis and analysis of new hyperbolograph mechanism. In: Analele Universităţii «Constantin Brâncuşi» din Târgu-Jiu, Seria Inginerie, nr. 4/2012, pp 65–75

16. Popescu I, Sass L(2002) Noi mecanisme conicografe, cu elemente flexibile. In: Construcţia de maşini, Nr. 12/2002, pp. 51–53
17. Popescu I, Romanescu AE, Sass L (2018) Generating ellipses by using mechanisms relying on simple geometrical aspects. In: Confereng 2018, Analele Univ. "C. Brâncuşi", Târgu-Jiu, nr. 2/2018, pp 18–23
18. Reuleaux F (1963) The kinematics of machinery. Dover Publications, NY
19. Reuleaux F, Kennedy ABW (eds) (1876) Kinematics of machinery: outlines of a theory of machines. Macmillan and Co., London
20. Sass L, Popescu I (2000) La synthese geometrique exacte d'un mecanisme elipsographe et d'un mecanisme de linearite. In: The 8th symposium on mechanisms and mechanical transmissions with international participation, Timişoara, pp 301–306
21. Sass L, Popescu I (2002) Mecanisme conicografe originale. In: Construcţia de maşini, Nr. 12/2002, pp 48–50
22. Shigley JE, Uicker JJ (1995) Theory of machines and mechanisms. McGraw-Hill, New York
23. Taimina D (2004) Historical mechanisms for drawing curves. Cornell University. http://ecommons.cornell.edu/bitstream/1813/2718/1/2004-9.pdf
24. Teodorescu D, Teodorescu ŞD (1975) Culegere de probleme de geometrie superioară, Ed. Did. Ped., Bucureşti
25. Teodorescu N ş.a. (1984) Probleme din Gazeta Matematică. Editura Tehnica, Bucuresti

Chapter 5
Mechanisms for Generating Some Plane Mathematical Curves

Abstract Based on geometrical problems, the synthesis of original mechanisms that trace different mathematical curves is made. Then the structure is analyzed and the calculations are made for the positions of the mechanisms, successive positions, diagrams of the curve generating point coordinates, diagrams of the displacements for some sliders, and the traced curves. Further curves generated by these mechanisms are also studied for different initial input data. In this way, with two original mechanisms, the cissoids (of the circle and of the straight line) are traced, in some cases in particular generating ellipses, parabolas and hyperbolas. Next Berard's curve is studied, where a point of the mechanism traces a branch of the curve, and another point, the other branch. By modifying some dimensions, other curves generated by the mechanism are obtained. We also made the synthesis of a new mechanism that generates egg-shaped curve. Analysis of the mechanism led to the generation of this curve. We changed some data of the mechanism resulting in similar or modified curves, but positioned in different trigonometric quadrants. Another original mechanism, also based on a geometry problem, describes double egg curve. By changing the initial data, similar curves are obtained, some of them being incomplete. Another original mechanism, based on a geometry problem, describes Bernoulli quartic. In this case, an additional kinematic chain is used to provide the midpoint of a straight segment of variable length at the movement of the mechanism. The chapter continues with a mechanism based also on geometrical principles, which traces Maclaurin's trisectrix. By changing some initial data, the mechanism traces completely different curves from the initial ones. There has also been a synthesis of some original mechanisms that trace ophiuride. In this case, other curves are obtained by modifying some dimensions of the mechanism. Finally, an original mechanism that traces the Pascal's snail is studied. Different snails are obtained by changing some dimensions of the mechanism.

I. Popescu et al., *Mechanisms for Generating Mathematical Curves*,
Springer Tracts in Mechanical Engineering,
https://doi.org/10.1007/978-3-030-42168-7_5

5.1 Mechanisms for Generating Cissoids

5.1.1 The Cissoid of a Circle

5.1.1.1 Introduction

On a straight line d, which is the system's ordinate, there are taken two fixed points O and Ω, and on the curve C, it is taken an M point (Fig. 5.1a–c). The locus of point P, where the $M\Omega$ secant intersects the perpendicular from O on OM is the cissoid of C curve [1]. The cissoid of a circle is called the cissoid of Diocles.

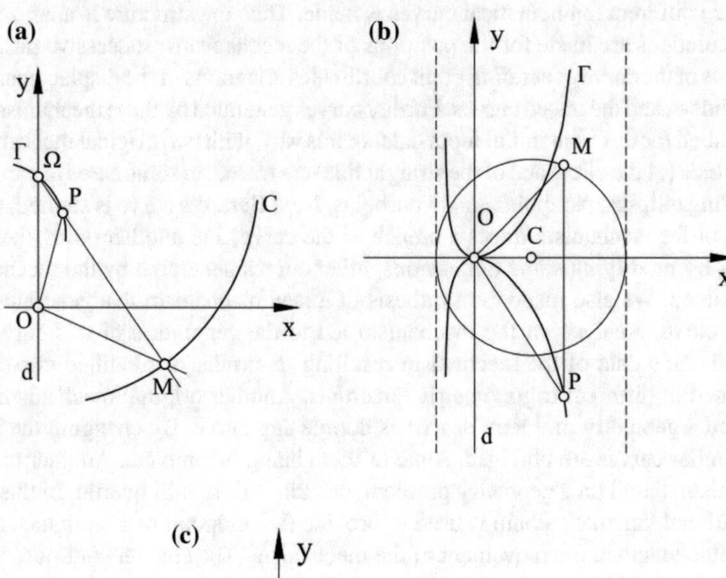

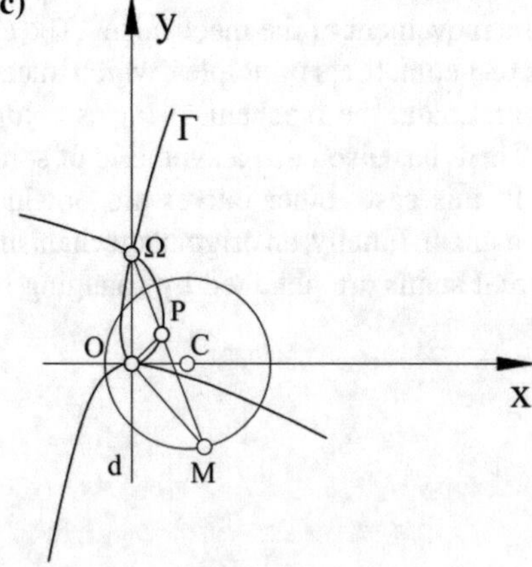

Fig. 5.1 Geometrical construction of the cissoid

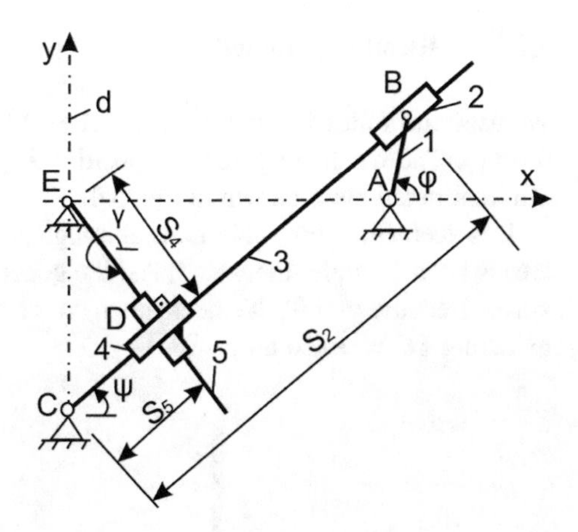

Fig. 5.2 Obtained mechanism

5.1.1.2 The Mechanism Synthesis

In Fig. 5.2, it is presented the mechanism which is constructed based on Fig. 5.1 [2].

Curve C becomes the circle of *AB* radius; the MΩ straight line becomes the *CB* element with variable length, that is why at *B* was mounted slider 2.

Point *O* shown in Fig. 5.1 becomes *E* in Fig. 5.2 and the perpendicular from *E* to *CB* was materialized into two perpendicular welded sliders. The generating point of the cissoid is the point *D*.

5.1.1.3 The Mechanism Analysis

Based on Fig. 5.2, with the contours method, Eqs. (5.1)–(5.5) can be written:

$$x_B = AB\cos\varphi + x_A = S_2\cos\psi \tag{5.1}$$

$$y_B = AB\sin\varphi + y_A = y_C + S_2\sin\psi \tag{5.2}$$

$$x_D = S_5\cos\psi + x_C = S_4\cos\gamma + x_E \tag{5.3}$$

$$y_D = y_C + S_5\sin\psi = S_4\sin\gamma + y_E \tag{5.4}$$

$$\gamma = \psi + 270 \tag{5.5}$$

from which the unknowns S_2, S_4, S_5, ψ are calculated.

5.1.1.4 Results Obtained

We used the following initial data: $XA = 40$, $AB = 40$, $YC = -50$; $XC = \mathrm{XE} = 0$ (mm). The mechanism for one position is given in Fig. 5.3. The mechanism for different positions is shown in Fig. 5.4.

It is seen from Fig. 5.4a that, although *B* makes a complete rotation, the *EDC* area is situated under the *x*-axis, and the generating point *D* is not passing above the *x*-axis. Because of this, the mechanism draws only a branch of the cissoid (Fig. 5.5a), by taking $YC = -120$ mm.

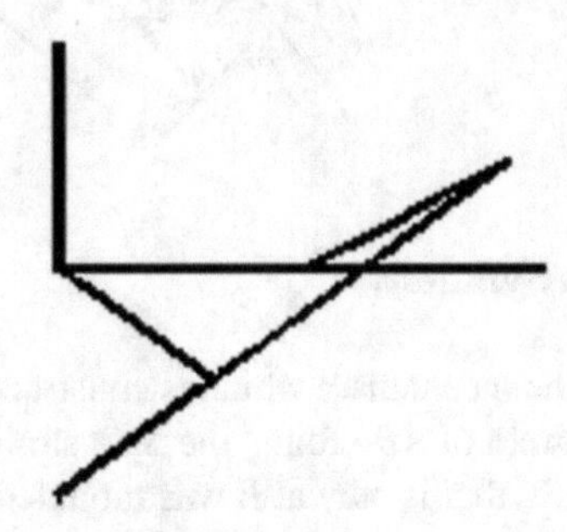

Fig. 5.3 Mechanism for a given position

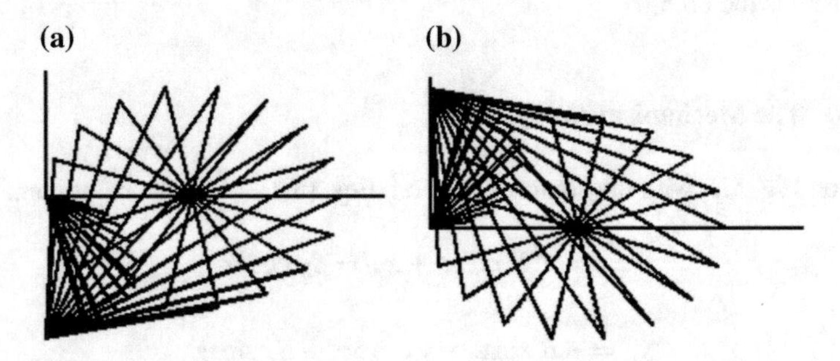

Fig. 5.4 Mechanism for different positions

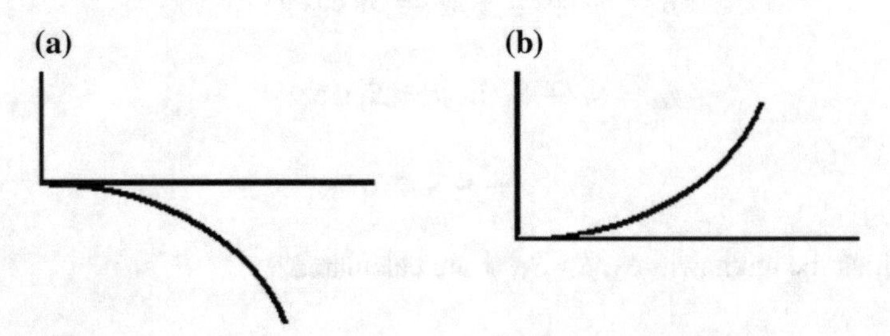

Fig. 5.5 Different branches of the cissoid

If it is assumed $YC = +120$ mm, then the other branch of the cissoid (Fig. 5.5b) is obtained. The successive positions are given in Fig. 5.4b.

5.1.2 The Straight Line Cissoid

5.1.2.1 Introduction

When the C curve of Fig. 5.1a is a straight line Δ, then the cissoid is a conic curve: an ellipse (Fig. 5.6a), a parabola (Fig. 5.6b), and a hyperbola (Fig. 5.6c) [3]. The ellipse is obtained when M is on the right side of Δ, the parabola when M is on Δ, the hyperbola when M is between Δ and OY, respectively.

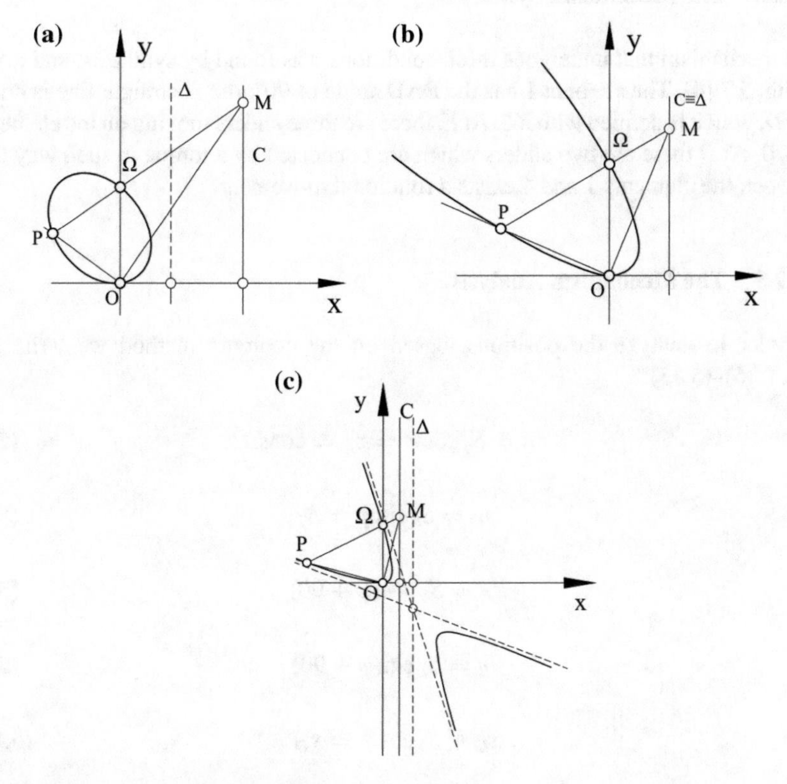

Fig. 5.6 Geometrical constructions of straight line cissoids

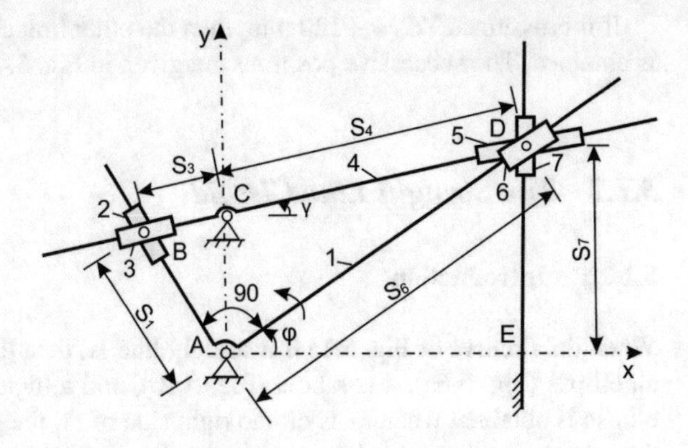

Fig. 5.7 Mechanism obtained

5.1.2.2 The Mechanism Synthesis

The mechanism that undergoes these conditions was found by synthesis, and given in Fig. 5.7 [4]. The element 1 has the BAD angle of 90°, the Δ straight line is noted as *ED*, which is defined with *XE*. At *E*, there are three sliders moving on the elements 4, 1, 0. At *B* there are two sliders which are connected by a torque in such way that between the elements 1 and 4 exists a rotational movement.

5.1.2.3 The Mechanism Analysis

In order to analyze the positions, based on the contours method we write the Eqs. (5.6)–(5.13):

$$x_D = S_6 \cos\varphi = x_E = \text{const.} \tag{5.6}$$

$$y_D = S_6 \sin\varphi = S_7 \tag{5.7}$$

$$x_B = S_1 \cos(\varphi + 90) \tag{5.8}$$

$$y_B = S_1 \sin(\varphi + 90) \tag{5.9}$$

$$x_C + S_4 \cos\gamma = x_D \tag{5.10}$$

$$y_C + S_4 \sin\gamma = y_D \tag{5.11}$$

$$x_B + S_3 \cos\gamma = x_C \tag{5.12}$$

$$y_B + S_3 \sin\gamma = y_C \tag{5.13}$$

5.1.2.4 Results Obtained

The values for YC and XE are given as input data. For the values of $YC = 50$ mm and $XE = 80$ mm the mechanism was obtained in one position, which is given in Fig. 5.8. The generating ellipse can be observed in Fig. 5.9 and the successive positions of the mechanism are given in Fig. 5.10.

Referring to Fig. 5.11, it is shown that XB varies by a sinusoid type curve, it also moves to the left of y-axis and YB is always positive.

For the computer program a jump instruction was provided for the cases when $S7$ advances above 200 mm, because it went beyond the drawing borders, which is

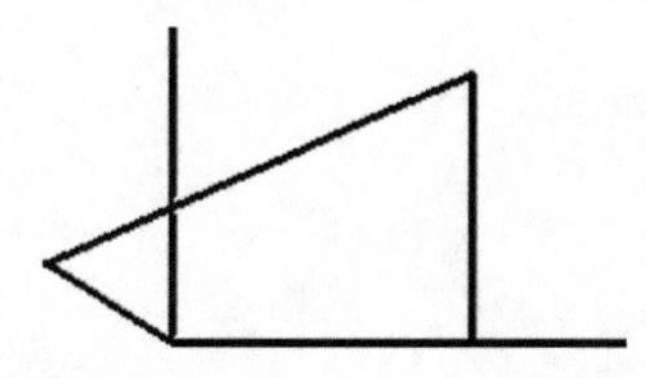

Fig. 5.8 Mechanism in one position

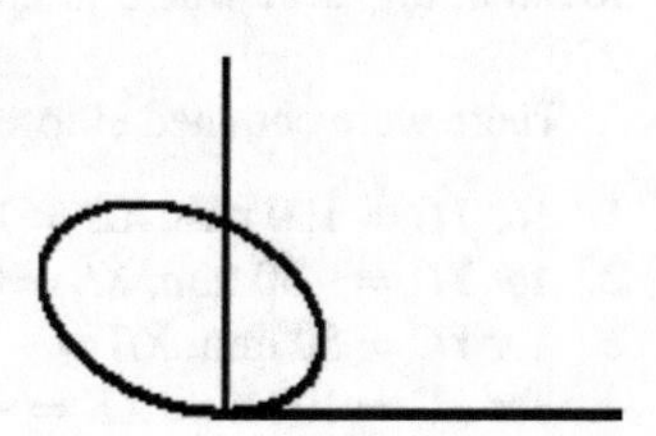

Fig. 5.9 Resulting ellipse

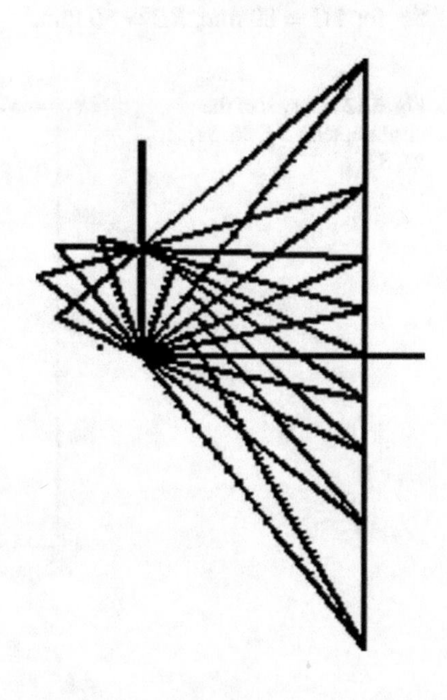

Fig. 5.10 Successive positions of the mechanism

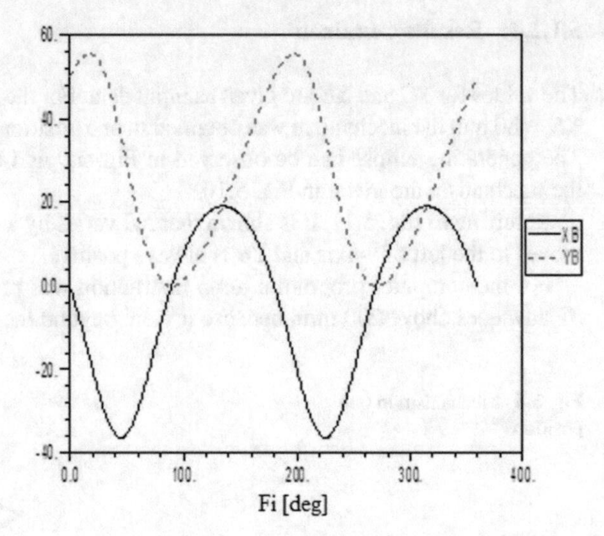

Fig. 5.11 Variations of the coordinates y_B, x_B

found in Fig. 5.12 where jumps occur at $S7$ and $S6$, for φ around values of 90° and 270°.

There were obtained ellipses (Fig. 5.13) for other sets of values, as follows:

1. for $YC = 100$ mm, $XE = 100$ mm;
2. for $YC = -50$ mm, $XE = 50$ mm;
3. for $YC = 50$ mm, $XE = -50$ mm;
4. for $YC = 100$ mm, $XE = -100$ mm;
5. for $YC = 80$ mm, $XE = 50$ mm.

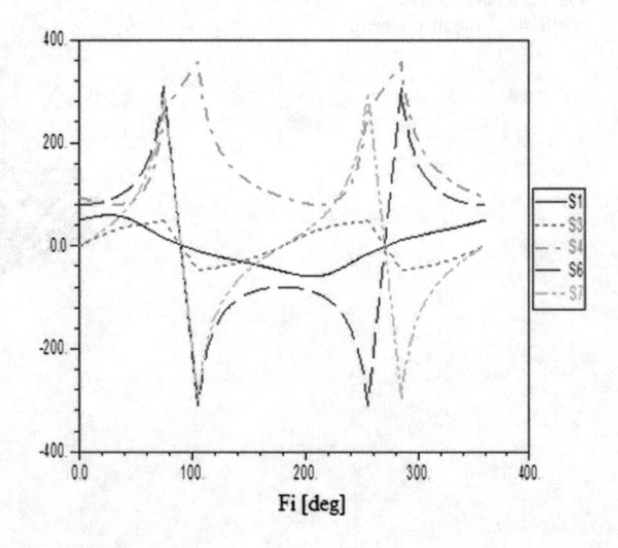

Fig. 5.12 Curves of the displacements $S4$, $S6$, $S1$, $S7$, $S3$

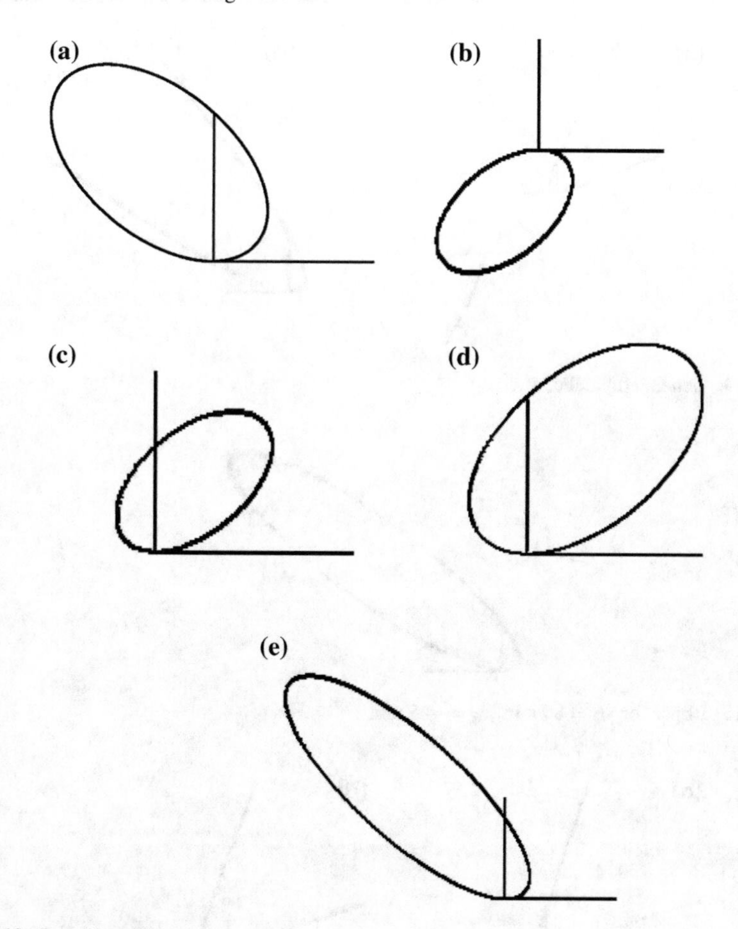

Fig. 5.13 Resulting ellipses

Changing the initial data led to resizing the ellipses dimensions and their axis orientation. When $YC < 0$, the ellipse is in quadrant III, and for $XE < 0$, the axis of the ellipse is placed in quadrant I.

We have also obtained the parabolas in Fig. 5.14:

1. for $YC = 100$ mm, $XE = 40$ mm;
2. for $YC = 100$ mm, $XE = -50$ mm.

For $YC = 80$ mm, $XE = -45$ mm, it resulted the ellipse in Fig. 5.15

In Fig. 5.16 there are presented the generating hyperbolas, for the values:

1. $YC = 100$ mm, $XE = 30$ mm;
2. $YC = 80$ mm, $XE = 25$ mm.

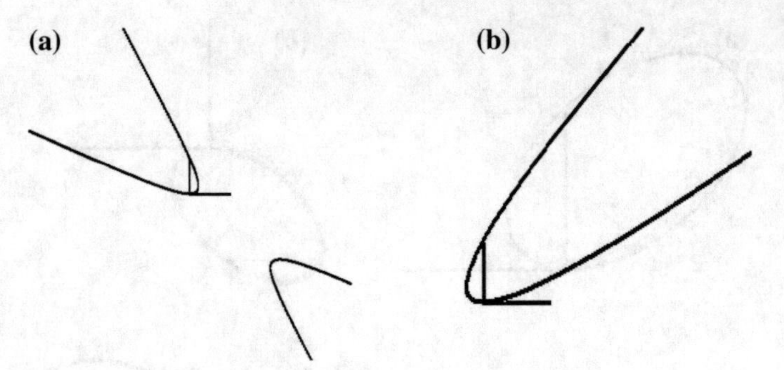

Fig. 5.14 Generated parabolas

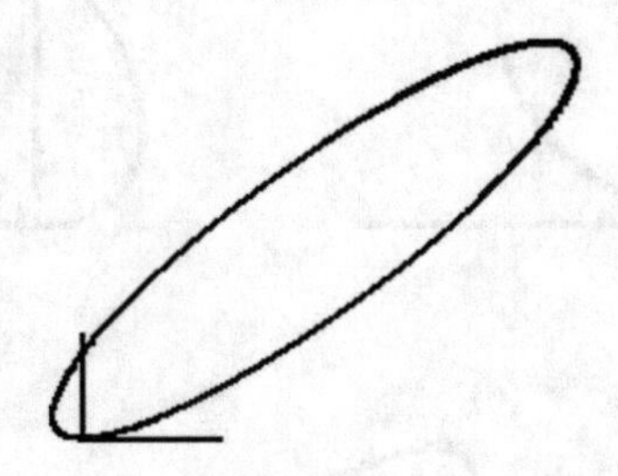

Fig. 5.15 Elipse for $y_C = 80$ mm, $x_E = -45$ mm

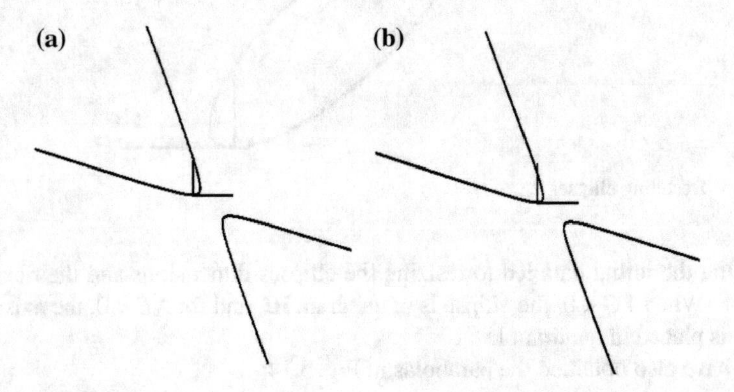

Fig. 5.16 Generated hyperbolas

5.2 Mechanism for Generating Berard's Curve

5.2.1 The Mechanism Synthesis

The rod curves of the rod-crank mechanism were studied by mathematicians since the invention of the steam engine [5]. Thus, in 1820, Bérard obtained such curves.

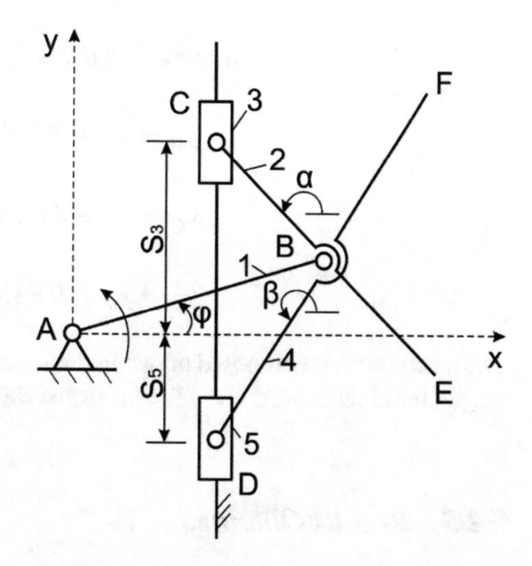

Fig. 5.17 Mechanism obtained

Bérard has obtained and defined them as locus of points which are situated on a straight line that goes from one end on a circle and the other end on another straight line. In [6] are given the equations and the particular cases of these curves. The movement of some straight lines is also designed, obtaining particular curves.

On the basis of these geometric considerations, the mechanism of Fig. 5.17 was conceived, which, by the E and F points, describes trajectories which are connected to each other, forming the curves of Berard. Structurally, the mechanism is R-RRP-RRP [7].

5.2.2 *The Mechanism Analysis*

On the basis of Fig. 5.17, Eqs. (5.14)–(5.23) are written:

$$x_B = AB\ \cos\varphi \tag{5.14}$$

$$y_B = AB\ \sin\varphi \tag{5.15}$$

$$x_C = AB\ \cos\varphi + BC\cos\alpha = \text{const.} \tag{5.16}$$

$$y_C = AB\ \sin\varphi + BC\sin\alpha = S_3 \tag{5.17}$$

$$x_E = x_B + BE\cos(\alpha + \pi) \tag{5.18}$$

$$y_E = y_B + BE\sin(\alpha + \pi) \tag{5.19}$$

$$x_D = x_B + BD\cos\beta = x_C = \text{const.} \tag{5.20}$$

$$y_D = y_B + BD\sin\beta = S_5 \tag{5.21}$$

$$x_F = x_B + BF\cos(\beta - \pi) \tag{5.22}$$

$$y_F = y_B + BF\sin(\beta - \pi) \tag{5.23}$$

The mechanism is composed of two link mechanisms with common crank, respecting the geometrical conditions: $AB = a$; $BC = BE = BD = BF = b$ şi $x_C = x_D < a + b$.

5.2.3 *Results Obtained*

5.2.3.1 Bérard's Curve

Bérard's curve is obtained for the particular case when $AB + x_C = BC$. For the values: $AB = 50$ mm, $BC = 75$ mm, $x_C = 25$ mm, the mechanism position in Fig. 5.18 was obtained and the successive positions are given in Fig. 5.19.

In Fig. 5.20, it is shown Bérard's curve, having $AB + x_C = BC$. The point F describes half of this curve, and the point E the other half, so that ultimately, we obtain a continuous curve.

In Fig. 5.21, there are shown the branches of Berard's curve:

1. described by the point E
2. described by the point F.

Specific to this mechanism is precisely that these branches are merged into a single curve—Bérard's curve.

Variations of $S3$ and $S5$ displacements are shown in Fig. 5.22.

The curves seem to have forms in extension without a jump. In reality, they have inverted symmetry, and they also have one jump at $\varphi = 180°$.

Fig. 5.18 Mechanism in a position

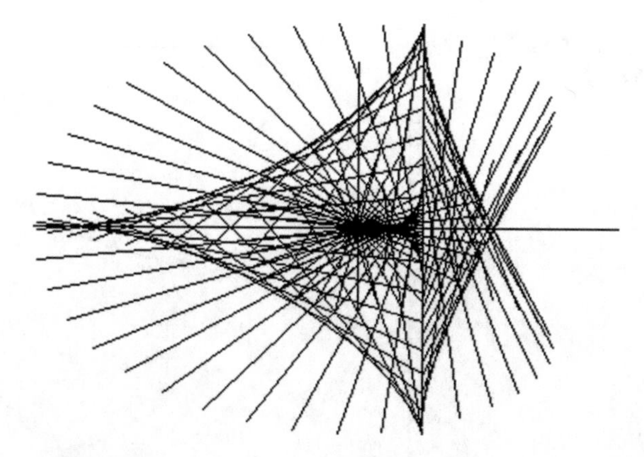

Fig. 5.19 Successive positions

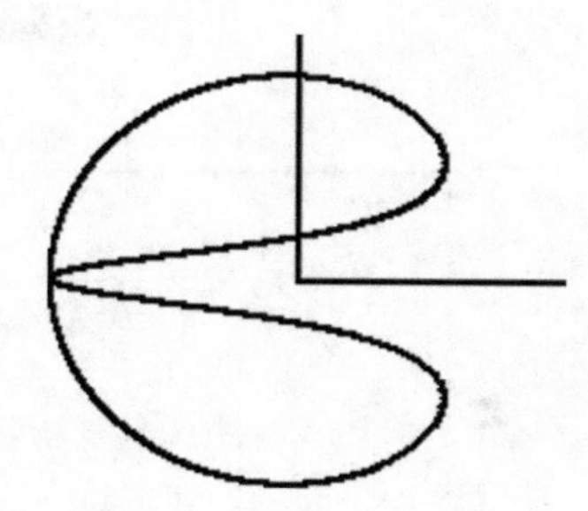

Fig. 5.20 Berard's curve

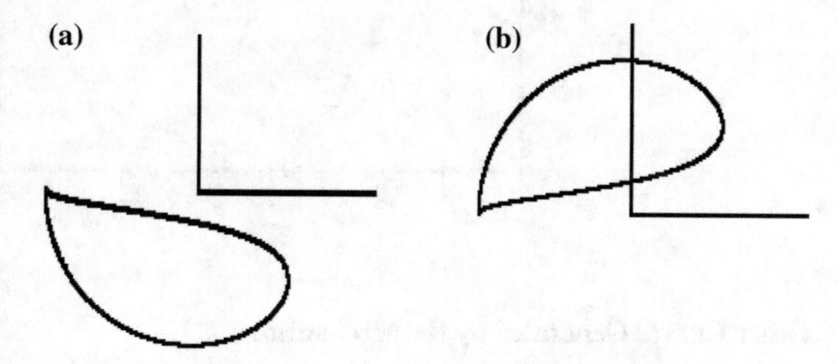

Fig. 5.21 Berard's curve branches

In Fig. 5.23, the variations of the coordinates of points E and F are shown.

It is noted that the abscissas of the points F and E have the same variation curve, and the other curves are similar to those of Fig. 5.22.

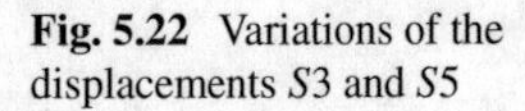

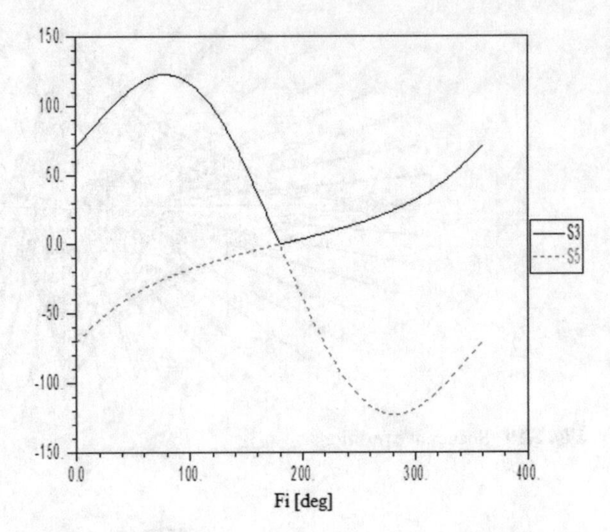

Fig. 5.22 Variations of the displacements $S3$ and $S5$

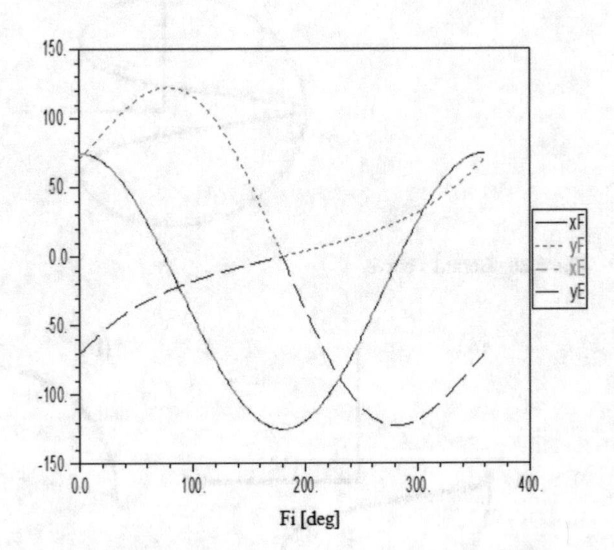

Fig. 5.23 Variations of the coordinates of E and F

5.2.3.2 Other Curves Generated by the Mechanism

There were also determined other curves generated by this mechanism. Thus, for $a = 50$ mm, $b = 50$ mm, $x_C = 0$, it resulted: one position in Fig. 5.24, the successive positions in Fig. 5.25 and the diagrams of Fig. 5.26.

The trajectory is an ellipse, and for the successive positions, a hypocycloid which resulted as an envelope can also be observed.

Referring to Fig. 5.26, it results that the abscissas of E and F have the same curve variation and the ordinates have symmetries and linear areas. Having $y_C = S3$ and $y_D = S5$, the curves for $S3$ and $S5$ were no longer given.

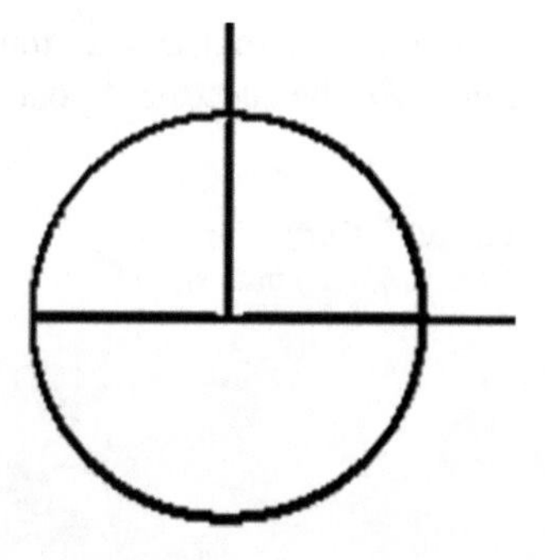

Fig. 5.24 Generated circle

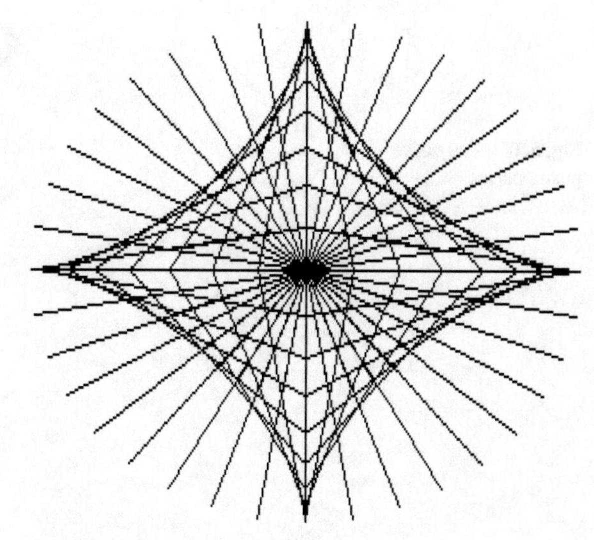

Fig. 5.25 Successive positions

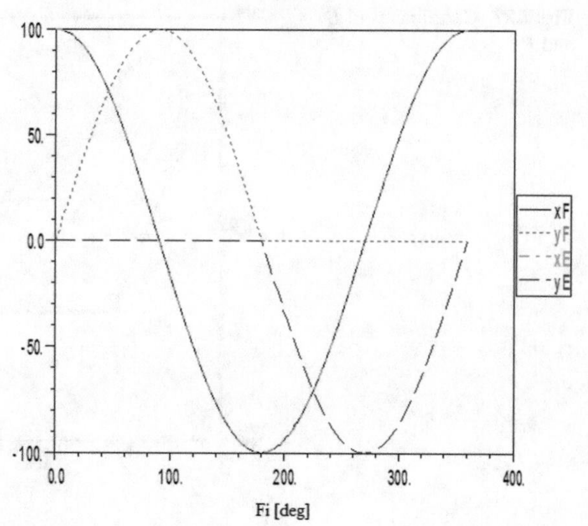

Fig. 5.26 Variations of the coordinates of E and F

For $a = 60$ mm, $b = 20$ mm, $x_C = 0$, it resulted: the trajectory which is shown in Fig. 5.27, the successive positions shown in Fig. 5.28 and the diagram of Fig. 5.29.

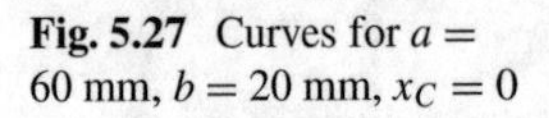

Fig. 5.27 Curves for $a =$ 60 mm, $b = 20$ mm, $x_C = 0$

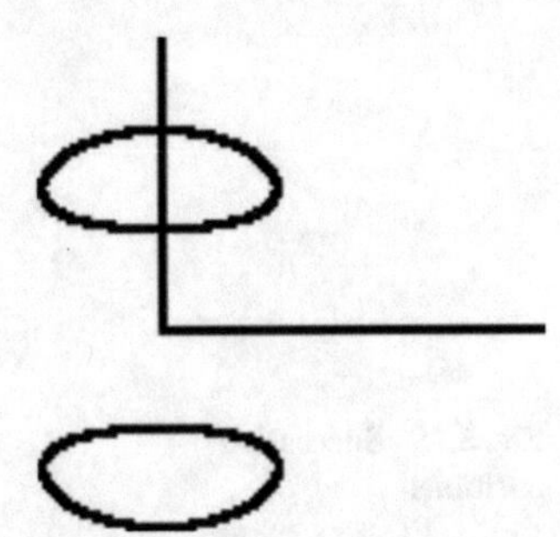

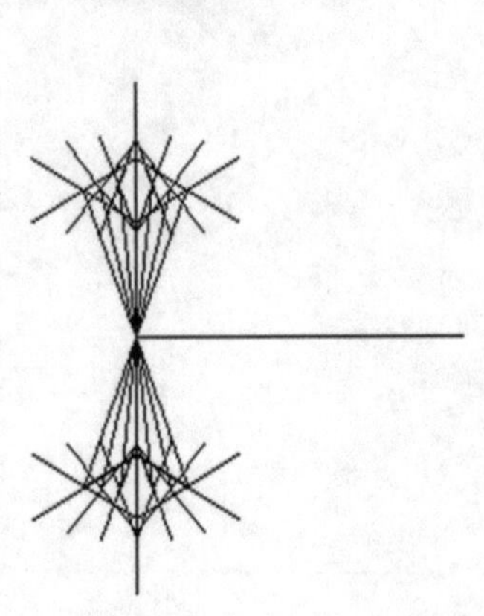

Fig. 5.28 Successive positions

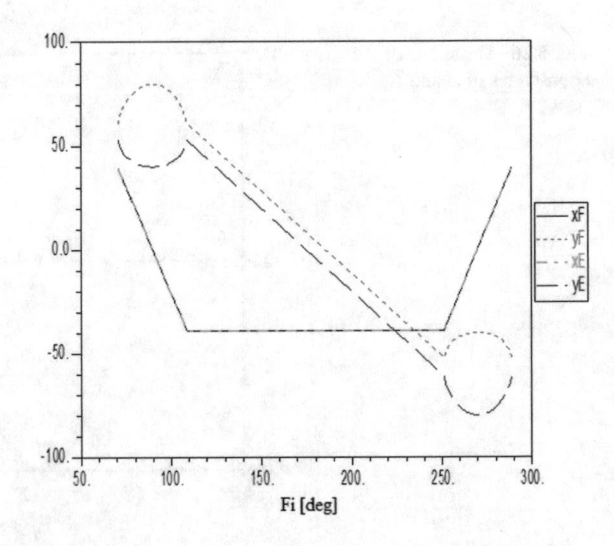

Fig. 5.29 Coordinates of E and F

Referring to Figs. 5.28 and 5.29, it can be observed that the mechanism only works in two small subdomains of the crank angle variation. Thus, two closed curves (Fig. 5.27) are obtained, which are symmetrical but are not connected to each other.

Referring to Fig. 5.29, it can be seen that the curves of the abscissa of the tracing points overlap and the ordinates have symmetrical variations, with values which are situated only in the two subintervals.

For equal values: $a = b = x_C = 40$ mm it resulted: the trajectory in Fig. 5.30, the successive positions shown in Fig. 5.31, and the diagram of Fig. 5.32.

It is noted that the trajectory seems to be a lemniscate (Fig. 5.30) and the successive positions (Fig. 5.31) do not exist in the range of 90°–270° of crank angle.

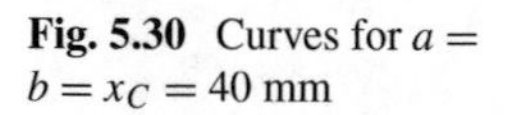
Fig. 5.30 Curves for $a = b = x_C = 40$ mm

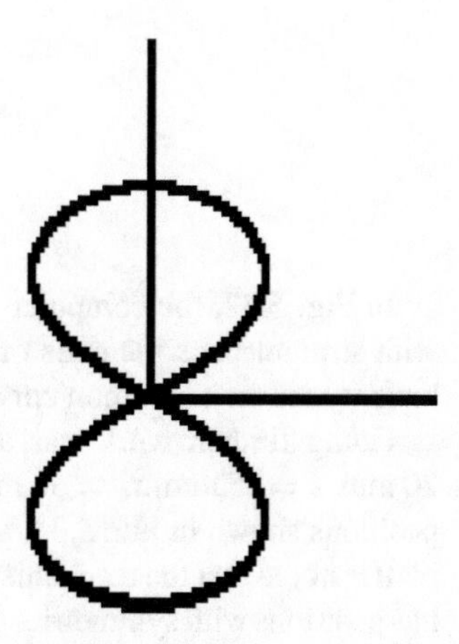

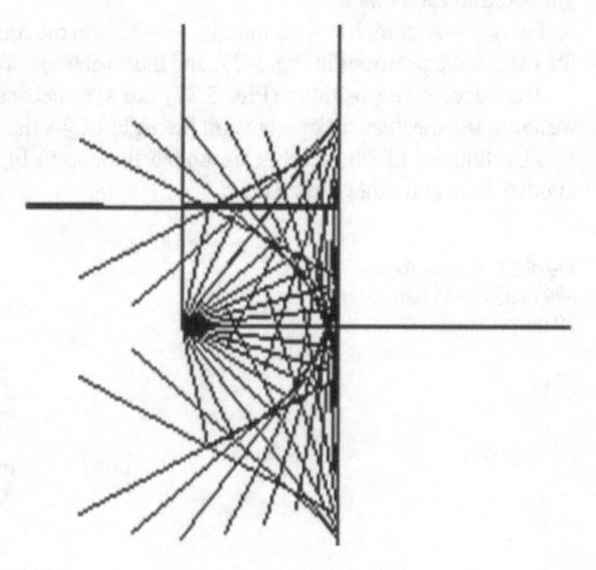

Fig. 5.31 Successive positions

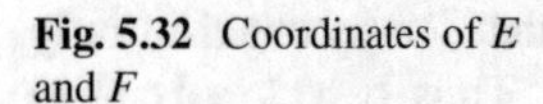

Fig. 5.32 Coordinates of E and F

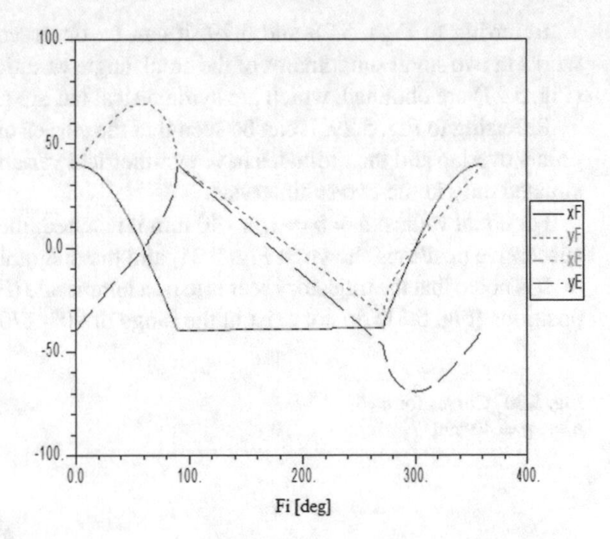

In Fig. 5.32, the computer program made the movement continuous by joining with straight lines the ends of the mentioned subinterval. Here, the abscissas also have values on a common curve.

Going further, work was accomplished with the following input data: $a = -20$ mm, $b = 55$ mm, $x_C = 30$ mm, to obtain the trajectory of Fig. 5.33, the successive positions shown in Fig. 5.34, and the diagram of Fig. 5.35.

It is noted that the mechanism works throughout the entire cycle (Fig. 5.35), and it has positions with symmetries (Fig. 5.34), but the trajectories (Fig. 5.33) are separate, unrelated to each other.

For $a = -20$ mm, $b = -55$ mm, $x_C = -30$ mm the trajectory resulted in Fig. 5.36, the successive positions in Fig. 5.37, and the variations of the coordinates in Fig. 5.38.

The successive positions (Fig. 5.37) are symmetrical to the ones in Fig. 5.34, meaning the mechanism operates on the right of y-axis.

The diagram of Fig. 5.38 is similar to the one in Fig. 5.35, but the values have another field and other dimensions.

Fig. 5.33 Curves for $a =$ –20 mm, $b = 55$ mm, $x_C =$ 30 mm

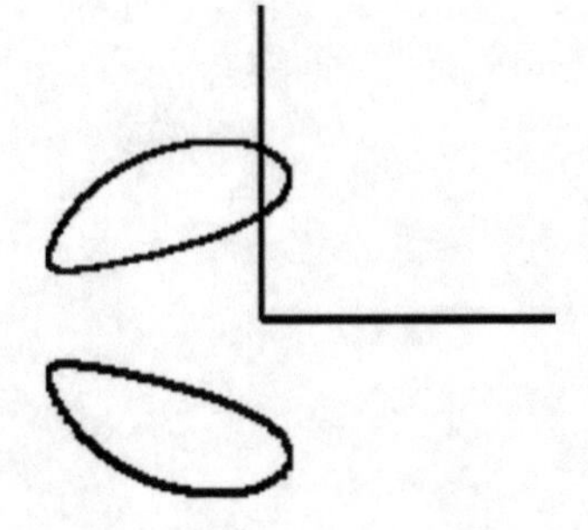

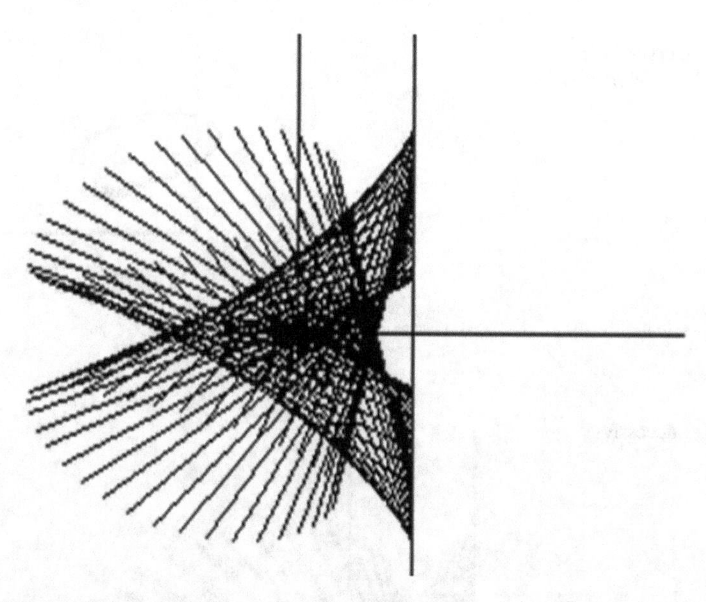

Fig. 5.34 Successive positions

Fig. 5.35 Coordinates of E and F

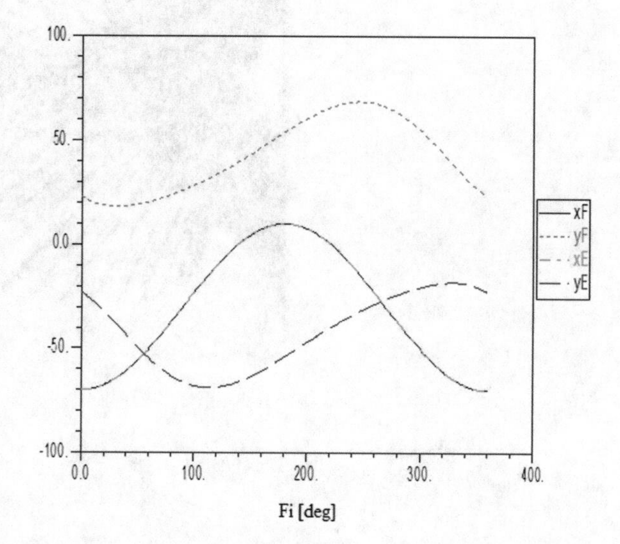

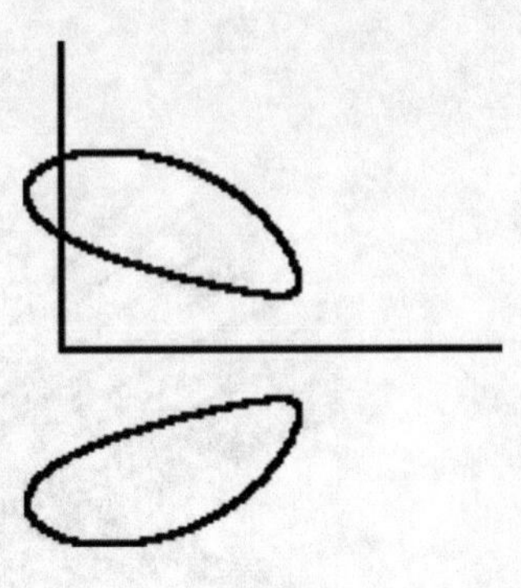

Fig. 5.36 Curves for $a = -20$ mm, $b = -55$ mm, $x_C = -30$ mm

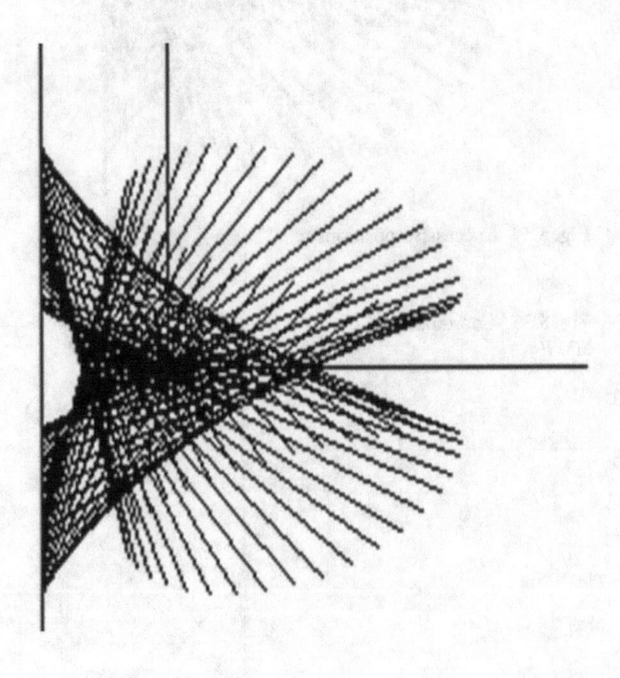

Fig. 5.37 Successive positions

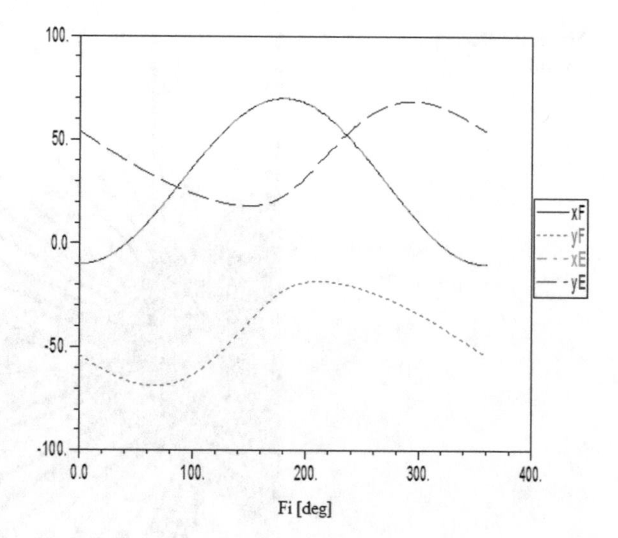

Fig. 5.38 Coordinates of E and F

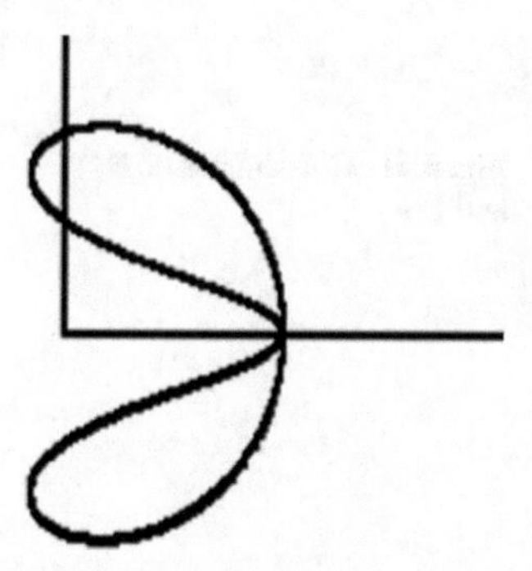

Fig. 5.39 Curve for a = 20 mm, b = 50 mm, x_C = –30 mm

For $a = 20$ mm, $b = 50$ mm, $x_C = -30$ mm, it results: the trajectory shown in Fig. 5.39, the successive positions shown in Fig. 5.40, and the diagram in Fig. 5.41.

In this case, Bérard's condition is fulfilled, meaning $a + abs\ (x_C) = b$, so that a Bérard continuous curve (Fig. 5.39) is obtained. The successive positions (Fig. 5.40) are symmetrical, and in the diagram (Fig. 5.41), the abscissas have the same variation curve and the ordinates have curves with symmetries.

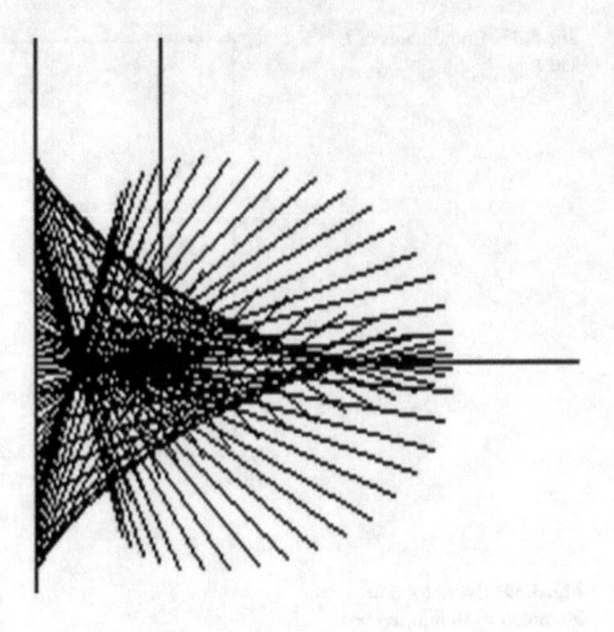

Fig. 5.40 Successive positions

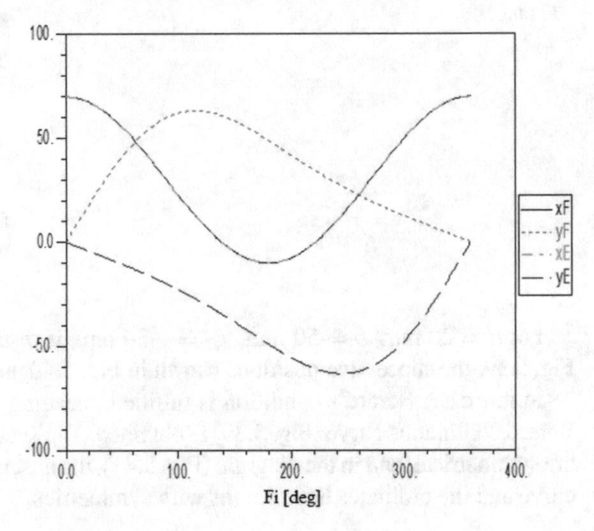

Fig. 5.41 Coordinates of E and F

5.3 Mechanism for Generating the Egg-Shaped Curve

5.3.1 Introduction

Among the curves which are used in practice is an egg-shaped curve. This has been geometrically studied. Geometric data of this curve and its correlation with other curves are given in [8].

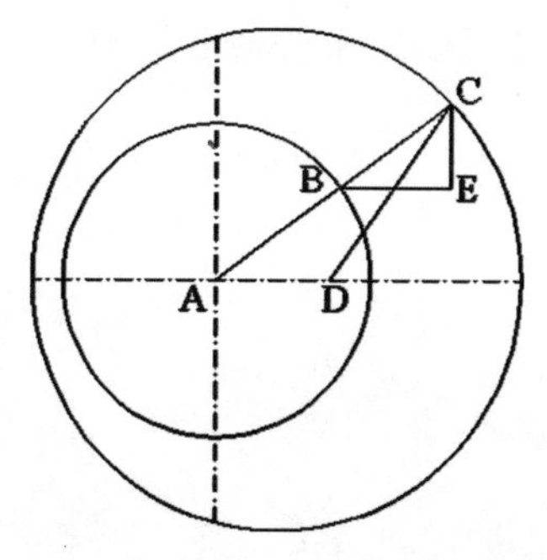

Fig. 5.42 Geometrical construction of the egg-shaped curve

In [9], it is considered a geometric solid of conoid type, having as profile in the frontal section the egg-shaped curve, obtaining different developed sections.

In [10], it is shown the geometrical construction in order to obtain this curve. There are also given examples of types of drawn curves. We consider a circle of AB radius and with center in A (Fig. 5.42) and another circle with DC radius, having the center in D. A parallel to the abscissa goes through point B and through C a parallel to the ordinate; they intersect in E. The locus of E, if ABC are collinear, is the egg-shaped curve.

The curve's equation is:

$$x^2 + y^2(1 + ax) = 1 \tag{5.24}$$

5.3.2 *The Mechanism Synthesis*

Next, an original mechanism that draws this curve is built and its possibilities in terms of paths generated are studied.

Based on previous remarks, the mechanism in Fig. 5.43 was designed. Since the length of BC is variable, a mechanism with oscillating $ABCD$ slider (Fig. 5.43) was initially designed [2].

Further, the straight lines BE and CE were traced. The abscissa of the point E is equal to the one of C and the ordinate is equal to the B ordinate (Fig. 5.43). Therefore, if the coordinates of C are determined, from the $ABCD$ oscillating slider mechanism it results the geometric position of the E tracing point. The element 5 with two sliders defines the position of E, but it is not guaranteed the vertical position of the element 4. The developed computer program managed to obtain the desired curve, but this results in only geometrical reasons i.e., the positions of points C and B define the coordinates of E. Structurally, the mechanism of Fig. 5.44 has the degree of mobility: $M = 3.6 - 2.8 = 2$, which makes it unusable.

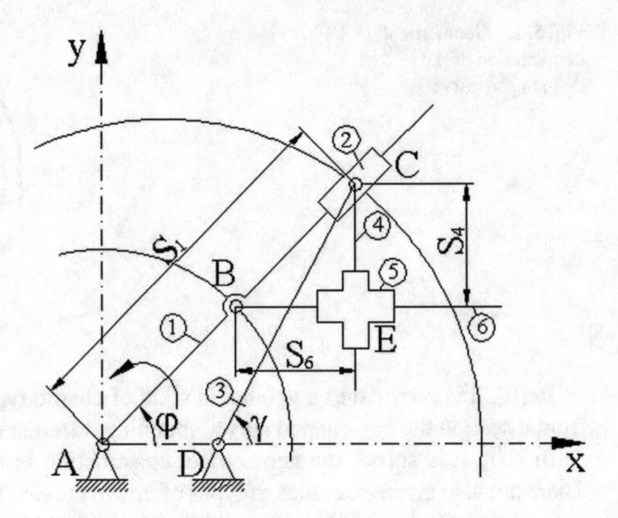

Fig. 5.43 Mechanism obtained

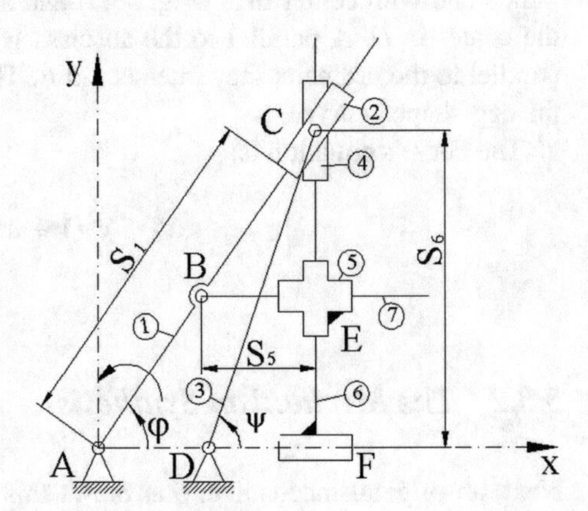

Fig. 5.44 Mechanism with $M = 2$

It is therefore necessary to specify the vertical position of element 4, which is achieved through the kinematic diagram of Fig. 5.44, where the element 4 was introduced, meaning a slider on 6, and 6 was linked to the base through the slider at F.

5.3.3 The Mechanism Analysis

The mechanism has the following structure (Fig. 5.45): R-PRR-RPP-PPR.

The necessary relations for the kinematic analysis are Eqs. (5.25)–(5.31):

$$x_B = AB\cos\varphi \tag{5.25}$$

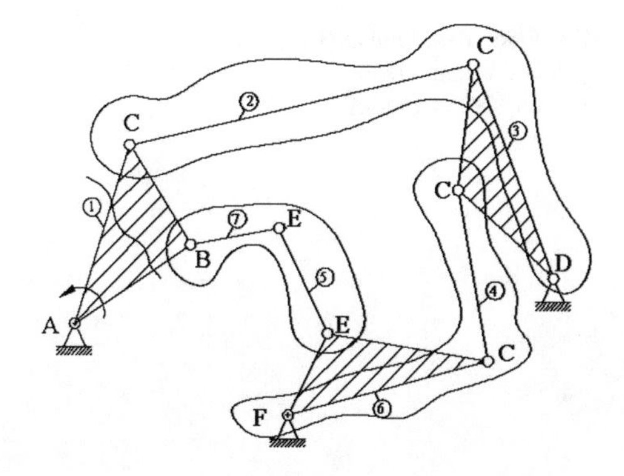

Fig. 5.45 Structure of the mechanism

$$y_B = AB \sin \varphi \tag{5.26}$$

$$x_C = x_D + DC \cos \psi = S_1 \cos \varphi \tag{5.27}$$

$$y_C = y_D + DC \sin \psi = S_1 \sin \varphi = S_6 \tag{5.28}$$

$$x_E = x_F = x_C \tag{5.29}$$

$$y_E = y_B \tag{5.30}$$

$$S_5 = x_C - x_E \tag{5.31}$$

To avoid discussions on the trigonometric quadrants, the calculation of $S1$ was made using Eq. (5.32):

$$DC^2 = x_D^2 + S_1^2 - 2x_D S_1 \cos \varphi \tag{5.32}$$

5.3.4 Results Obtained

In Fig. 5.46, the two circles with their centers at A and D are shown, the resulting curve, which is egg-shaped and the mechanism position at $\varphi = 50°$. It appears that the mechanism draws the desired curve.

In Fig. 5.47, the successive positions of the mechanism are shown, from which the positions of the BEC triangle are clearly observed, with sides of varying lengths.

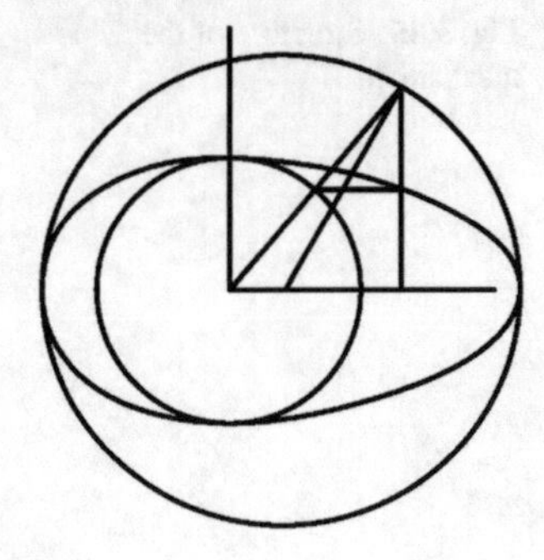

Fig. 5.46 Resulting curve ($AB = 25$ mm; $XD = 10$ mm; $DC = 45$ mm)

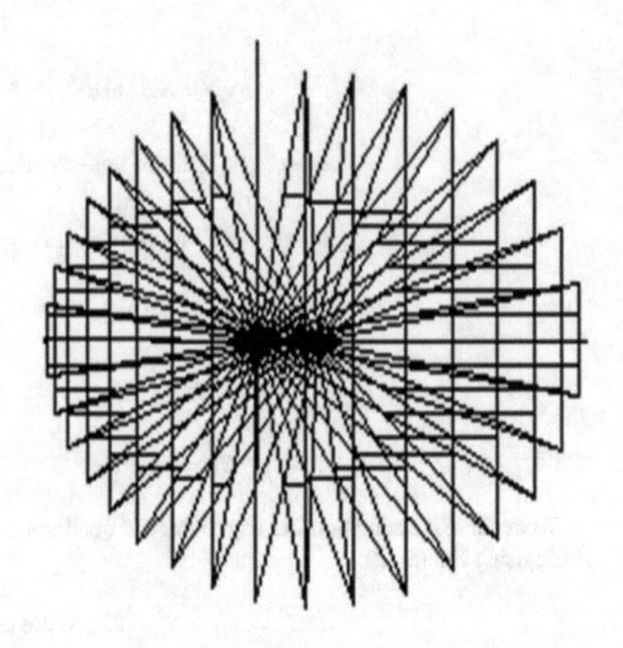

Fig. 5.47 Successive positions

In Fig. 5.48, the curves of *E* coordinates are shown. Each curve has a symmetry, reaching also the negative domain.

In Fig. 5.49, the curves for the displacements $S1$, $S5$, $S6$ are shown. These curves also have symmetries. The $S6$ curve is identical to that of *YE* from Fig. 5.48.

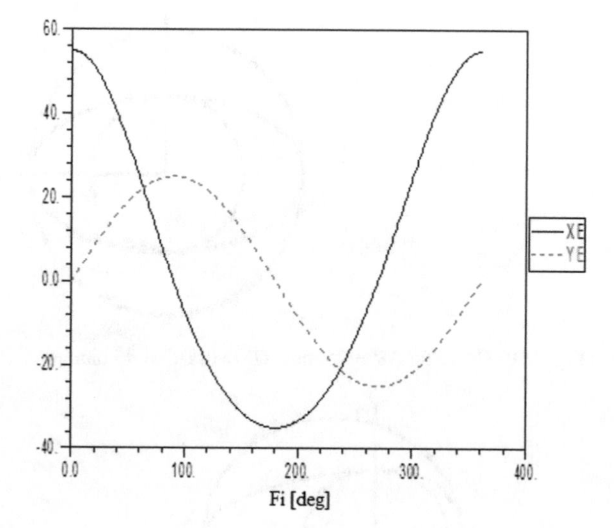

Fig. 5.48 Coordinates of E

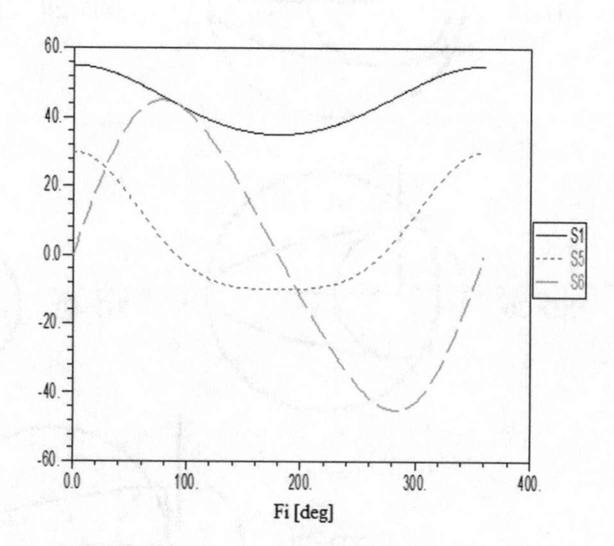

Fig. 5.49 Curves for $S1$, $S5$, $S6$

5.3.5 *The Influence of Inputs on the Curve's Shape*

The influence of XD size on the shape of the resulted curve was studied. For $XD = 0$, it was obtained the symmetrical "egg" (Fig. 5.50).

Increasing just XD ($AB = 25$ mm; $DC = 45$ mm), the "egg" becomes sharper at the right end, and the circle with center at D approaches the circle with center at A, then intersects it, the left area of the curve approaching in shape of a straight line segment (Fig. 5.51).

For $XD > DC$, the mechanism does not work for the entire cycle, there is a break, meaning a blocking in a subinterval (Fig. 5.51 $XD = 50$ mm $> DC = 45$ mm).

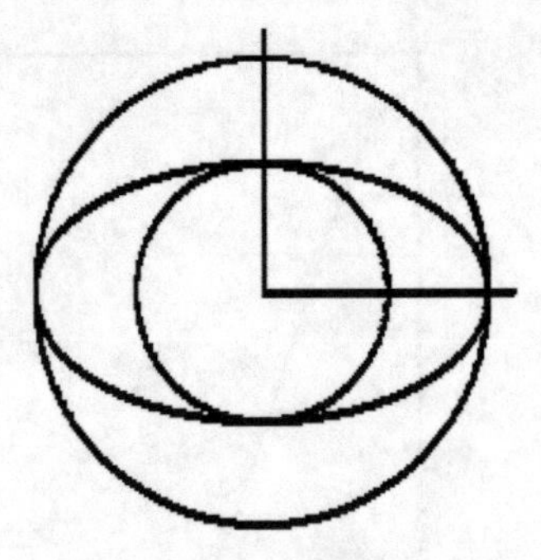

Fig. 5.50 Curve for $AB = 25$ mm, $XD = 0$, $DC = 45$ mm

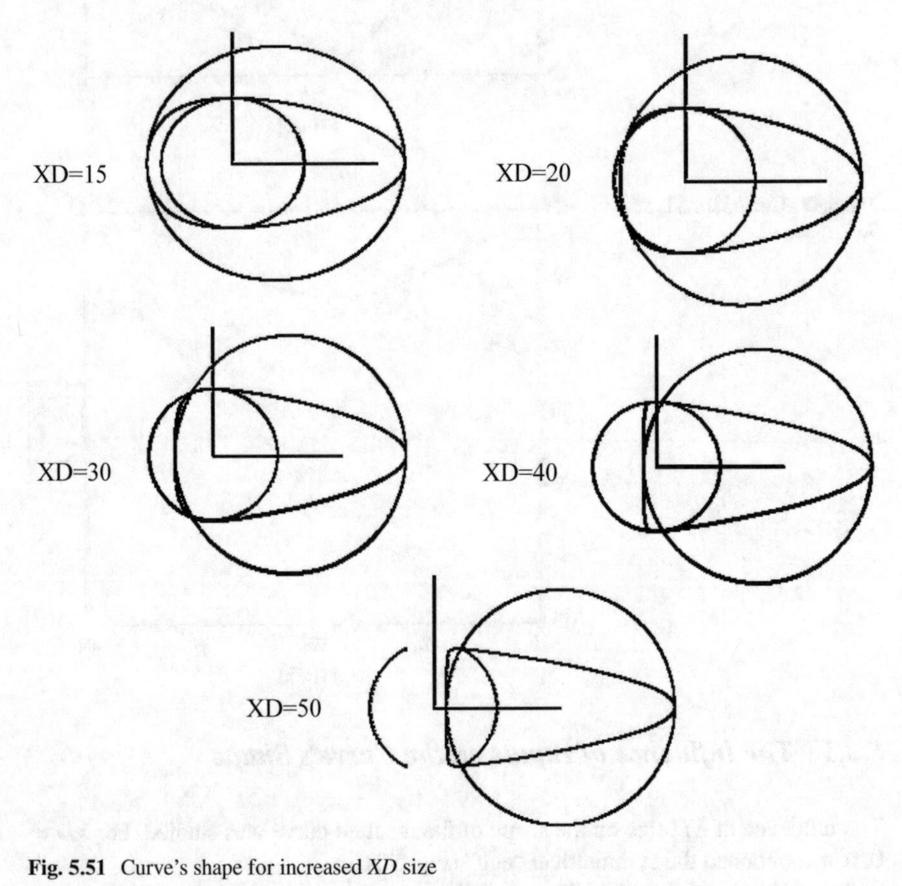

Fig. 5.51 Curve's shape for increased *XD* size

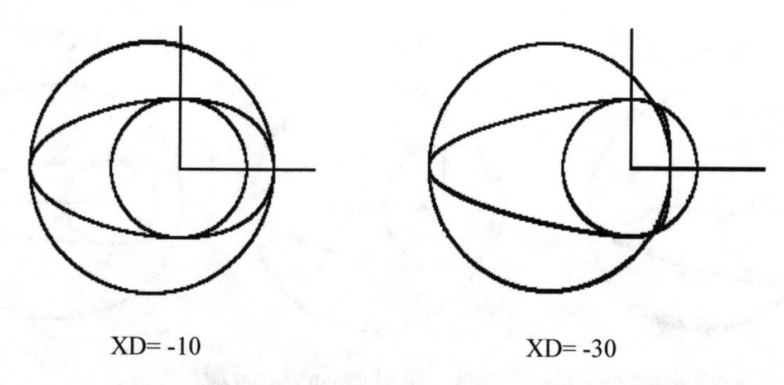

Fig. 5.52 Curve's shape for *XD* negative

If $XD < 0$, the above curves are obtained, but they are oriented symmetrically in relation to the ordinate (Fig. 5.52).

Next, the influence of the DC size on the curve's shape was studied.

It appears (Fig. 5.53) that for $DC < XD$ the mechanism does not work on a subinterval. For $DC = XD$, the drawn curve partially overlaps an arc of the circle with the center in *A* (Fig. 5.54).

By increasing *DC* ($XD = 30$ mm = const.), the curve becomes sharper at the right end and more flat-shaped at the left end (Fig. 5.55).

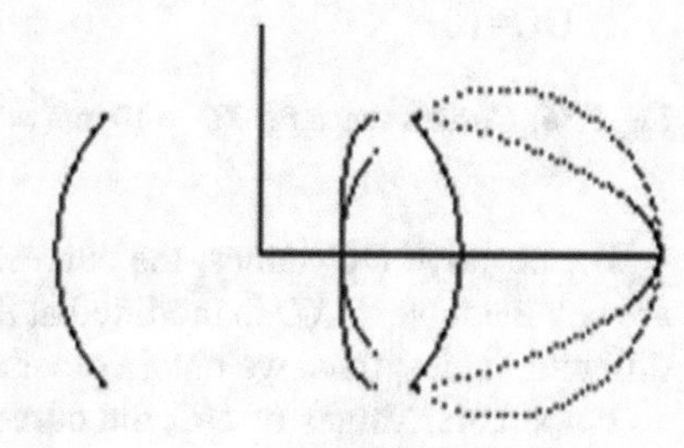

Fig. 5.53 Curve for $XD =$ 30 mm, $DC = 20$ mm

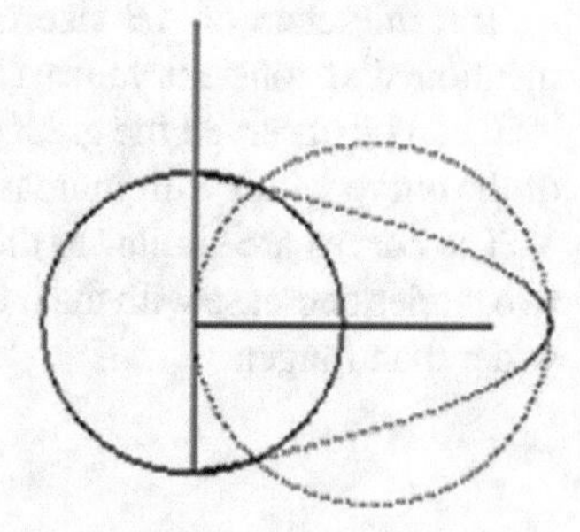

Fig. 5.54 Curve for $XD = DC = 30$ mm

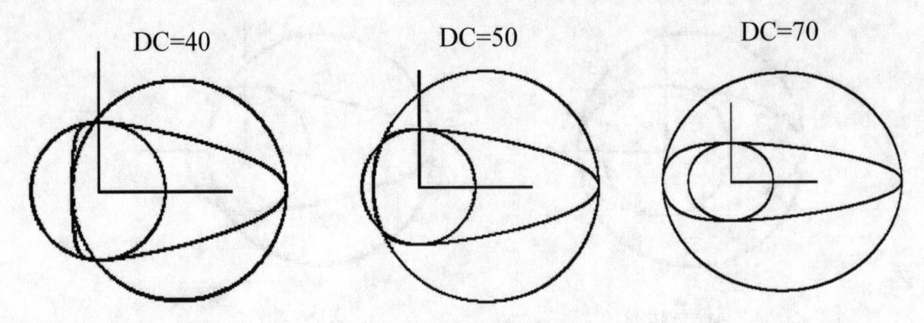

Fig. 5.55 Curve's shape for $XD = 30$ mm = const and increased DC

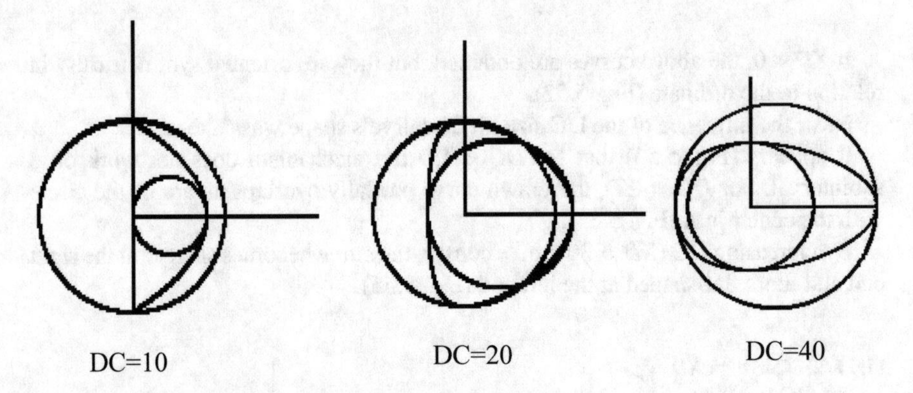

Fig. 5.56 Curve's shape for $XD = 10$ mm = const and increased DC

For negative *DC* values, the curves, which are similar to those for positive *DC*, are obtained. Next, *XD* is modified at the value of $XD = 10 =$ constant and *DC* has different values; thus, we obtain other curves (Fig. 5.56).

For smaller values of *DC*, the curves in Fig. 5.56 are different from the ones in Fig. 5.55, and for high values, the curves are similar to the ones in Fig. 5.55.

The influence of *AB* size on the shape of the curve, when the other sizes are mentioned at constant values ($XD = 10$ mm, $DC = 45$ mm), was also studied.

It can be observed the great difference between the two circles and the increasing of the curve width with increasing *AB* size in Fig. 5.57.

The curves are similar to the ones obtained above, but the distances between the two circles decrease with the increase of *AB*. At large values of *AB*, the curve becomes wider than longer.

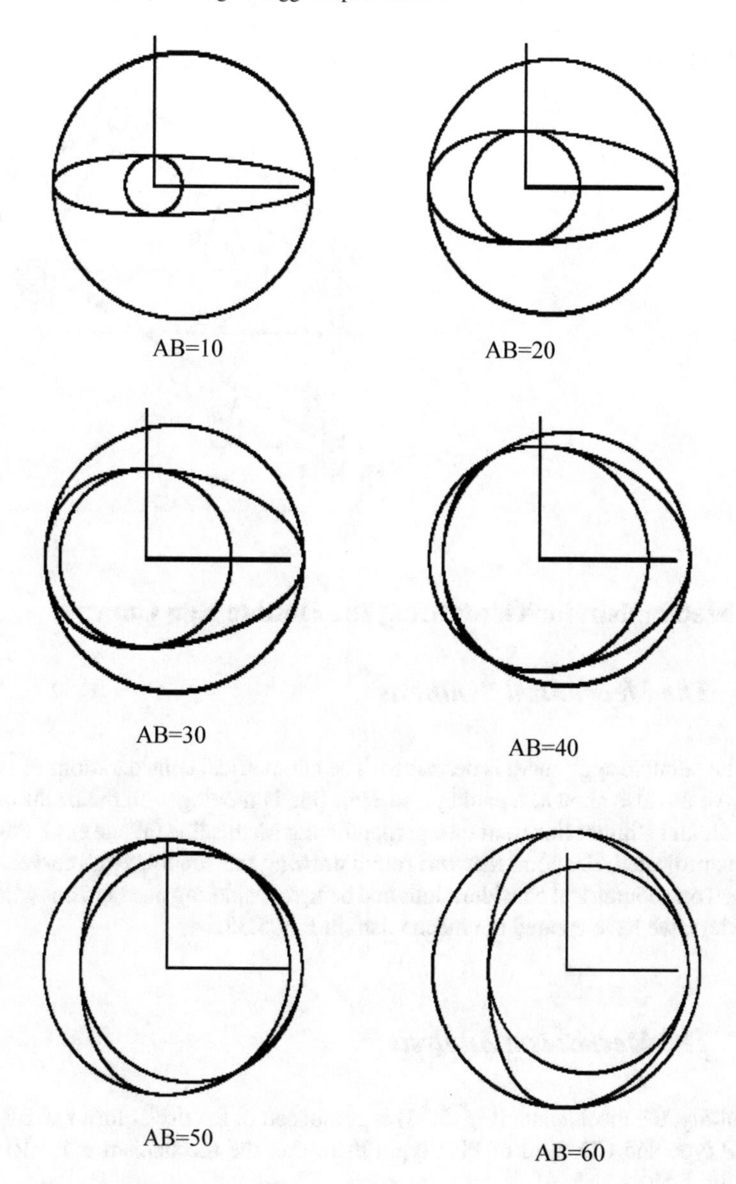

Fig. 5.57 Influence of *AB* size upon the curve's shape

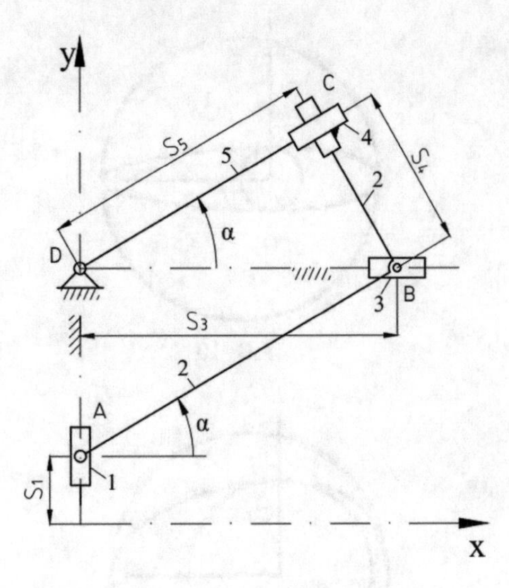

Fig. 5.58 Mechanism obtained

5.4 Mechanism for Generating the Double Egg Curve

5.4.1 The Mechanism Synthesis

In [5] the "double egg" curve is described. The geometrical considerations of tracing this curve are also shown, meaning a straight line is moving with the heads on two perpendicular straight lines and on a perpendicular on this line (at one end), the axis' origin is projected, the point that was found drawing the "double egg" curve.

Based on geometrical considerations and being considering mechanisms with bars and sliders, we have created the mechanism in Fig. 5.58.

5.4.2 The Mechanism Analysis

Structurally, the mechanism (Fig. 5.58) is composed of the driver link 1, ABB dyad of RRP type and CD dyad of PPR type; therefore, the mechanism is P-RRP-PPR type (Fig. 5.59).

Based on Fig. 5.58 and the closed-loop method, Eqs. (5.33)–(5.39) are written:

$$x_B = S_3 = x_A + AB\cos\alpha \tag{5.33}$$

$$y_B = S_1 + AB\sin\alpha = y_D \tag{5.34}$$

$$x_C = x_B + S_4\cos(\alpha + 90) = S_5\cos\alpha \tag{5.35}$$

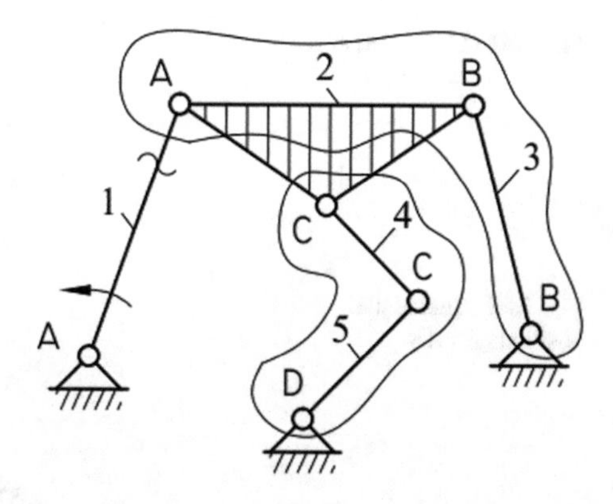

Fig. 5.59 Structure of the mechanism

$$y_C = y_B + S_4 \sin(\alpha + 90) = S_5 \sin\alpha + y_D \tag{5.36}$$

$$\tan\alpha = \frac{y_B - y_D + S_4 \cos\alpha}{x_B - S_4 \sin\alpha} \tag{5.37}$$

$$S_4 = \frac{y_B - y_D - x_B tg\,\alpha}{-\sin\alpha\tan\alpha - \cos\alpha} \tag{5.38}$$

$$S_5 = \frac{x_C}{\cos\alpha} \tag{5.39}$$

5.4.3 Results Obtained

In Fig. 5.60, the mechanism for $S1 = 20$ mm, $AB = 60$ mm and $YD = 30$ mm (values determined by testing) is shown. The axis system of Fig. 5.58 and the perpendicular axis on which are the sliding points A and B are noted. The BCD right angle is also noted.

In Fig. 5.61, the desired curve is shown; therefore, the mechanism is correct.

The successive positions of the AB rod, i.e., the straight line that slides with the ends on the axes, envelope an astroid (Fig. 5.62).

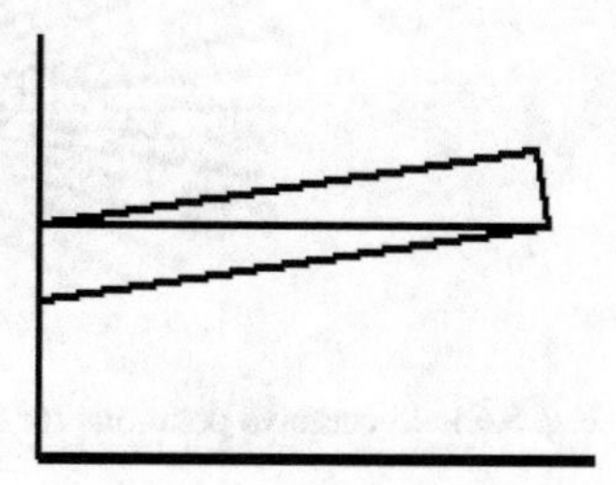

Fig. 5.60 Mechanism in a position

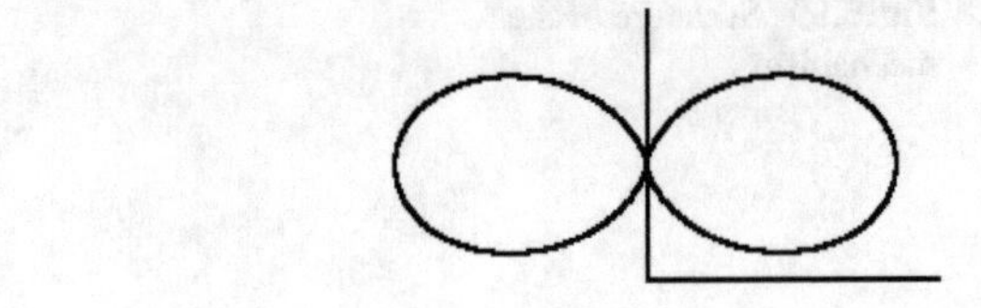

Fig. 5.61 Generated curve

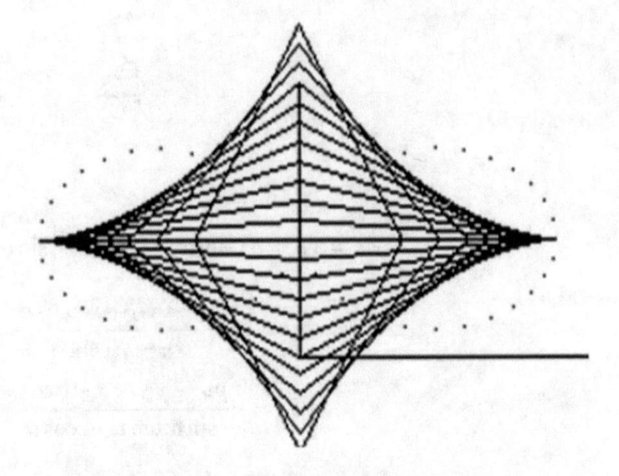

Fig. 5.62 Successive positions for *AB*

In Fig. 5.63, the successive positions of *DC* straight line with variable length are shown. The point *C* traces the double egg.

In Fig. 5.64, the successive positions of the *BC* straight line with variable length are shown. A symmetry relative to the ordinate is noted to appear.

The successive positions of the entire mechanism are shown in Fig. 5.65. It should be noted that the mechanism outlines a single "egg" from the "double egg," and if it is fitted symmetrically on the left y-axis, it also traces the other half of the curve. The developed computer program has $\pm$ in front of a square root, the solution with "+" providing the half of curve from the right, and the solution with the "−" providing the one on the left (Fig. 5.66).

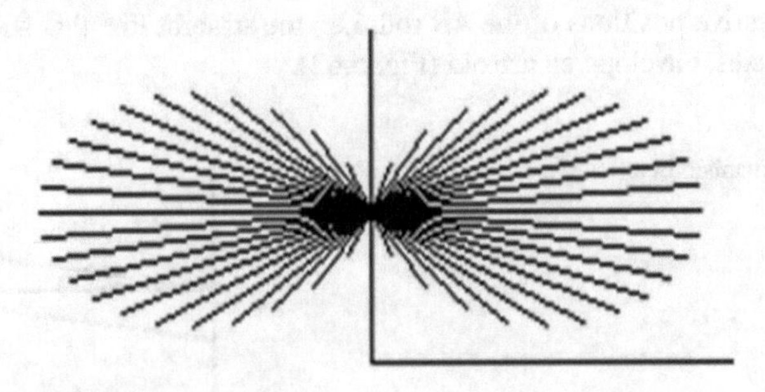

Fig. 5.63 Successive positions for *DC*

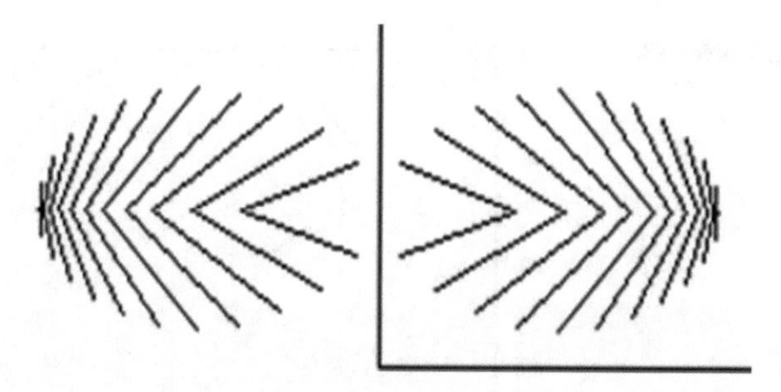

Fig. 5.64 Successive positions for *NC*

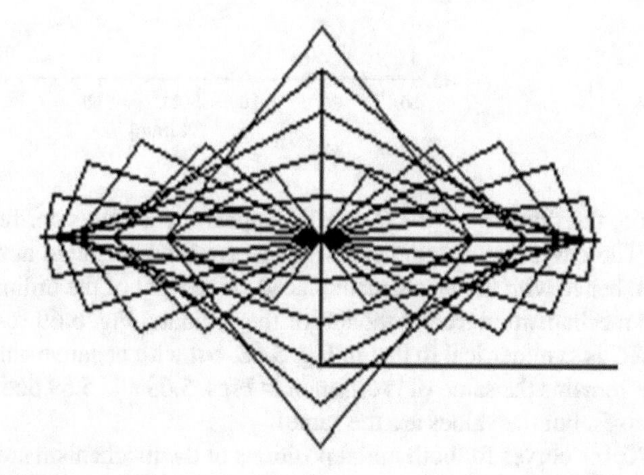

Fig. 5.65 Successive positions for the mechanism

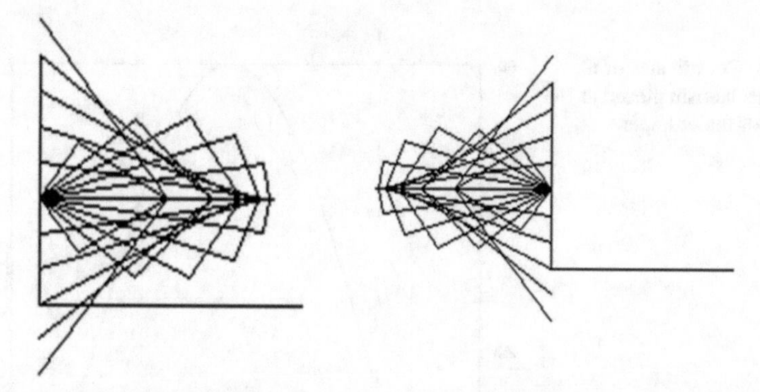

Fig. 5.66 Curve's right half and left half

In Fig. 5.67, the diagrams for *S*3 (upper curve), *S*5 (the intermediate one), and *S*4 (the bottom one) appear. It is noted that *S*4 has a sinusoidal variation and the other curves are similar to parabolas. There are no interruptions; therefore, the mechanism works a full cycle in the range of variation of *S*1 on the abscissa.

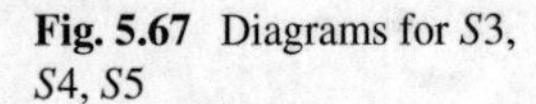
Fig. 5.67 Diagrams for $S3$, $S4$, $S5$

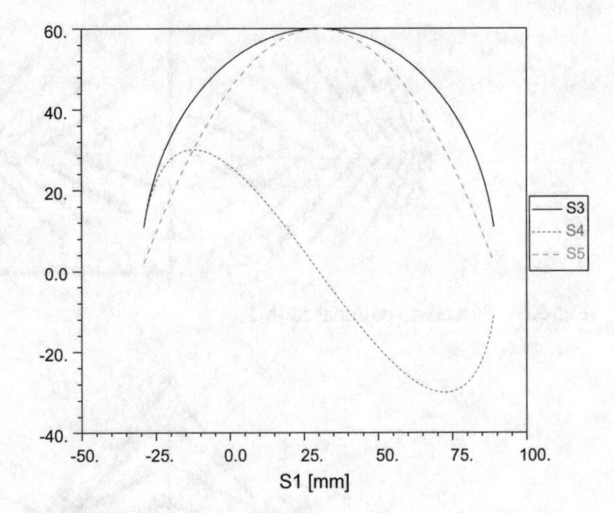

In Fig. 5.68, the diagrams for XC and YC appear, hence the coordinates of the tracer point. The curves have symmetries. There are those resulted at calculations with "+" sign, hence with the mechanism placed at the right of the ordinate. For "–" sign with the mechanism placed at the left of the ordinate, Fig. 5.69 results, where the curve of XC is symmetrical to that in Fig. 5.68, but with negative values for XC. The YC curve remains the same (it is different in Figs. 5.68 and 5.69 due to different scales of the axes, but the values are the same).

In Fig. 5.70, the curves for both initial positions of the mechanism are given. The parallels with the abscissa of the curves appear because of the transition from a set of values to other, the cycles starting from scratch.

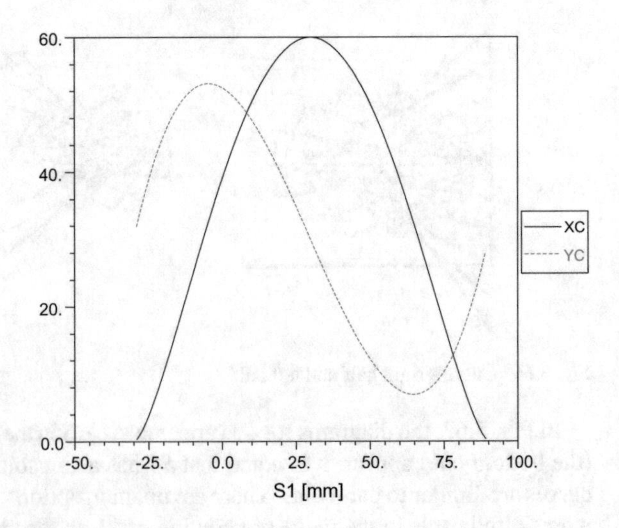

Fig. 5.68 Coordinates of C for the mechanism placed at the right of the ordinate

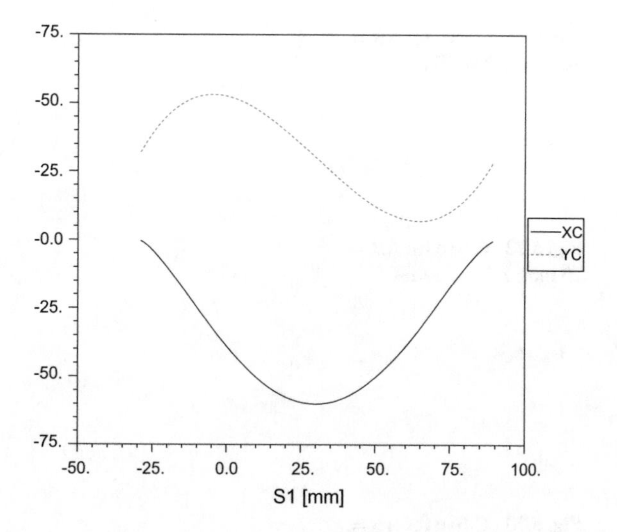

Fig. 5.69 Coordinates of *C* for the mechanism placed at the left of the ordinate

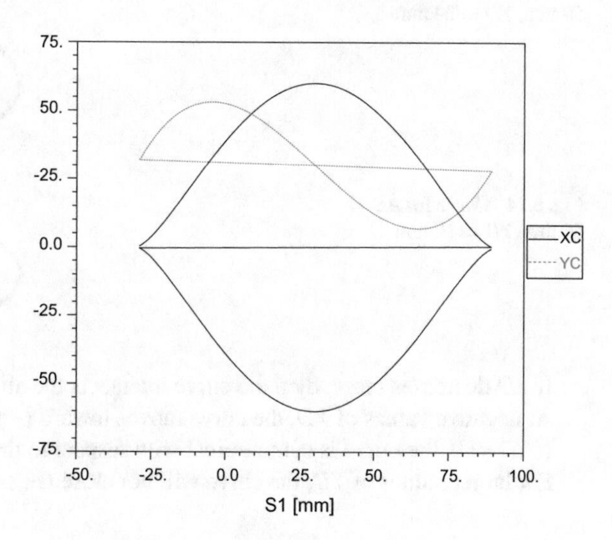

Fig. 5.70 Coordinates of *C* for both initial positions of the mechanism

5.4.4 The Influence of Inputs on the Curve's Shape

All the previous analyses were related to the case of $AB = 60$ mm, $YD = 30$ mm. We have tried other sets of values. Thus, if $AB > YD$ we obtain the desired curve, but positioned as in Fig. 5.71, meaning nearly tangent to the abscissa.

When $AB < YD$, the resulting curve is shifted higher than in the previous case, and the small value of AB causes small curve dimensions (Fig. 5.72).

In Fig. 5.73 is presented the case in which the curve is tangent to the abscissa.

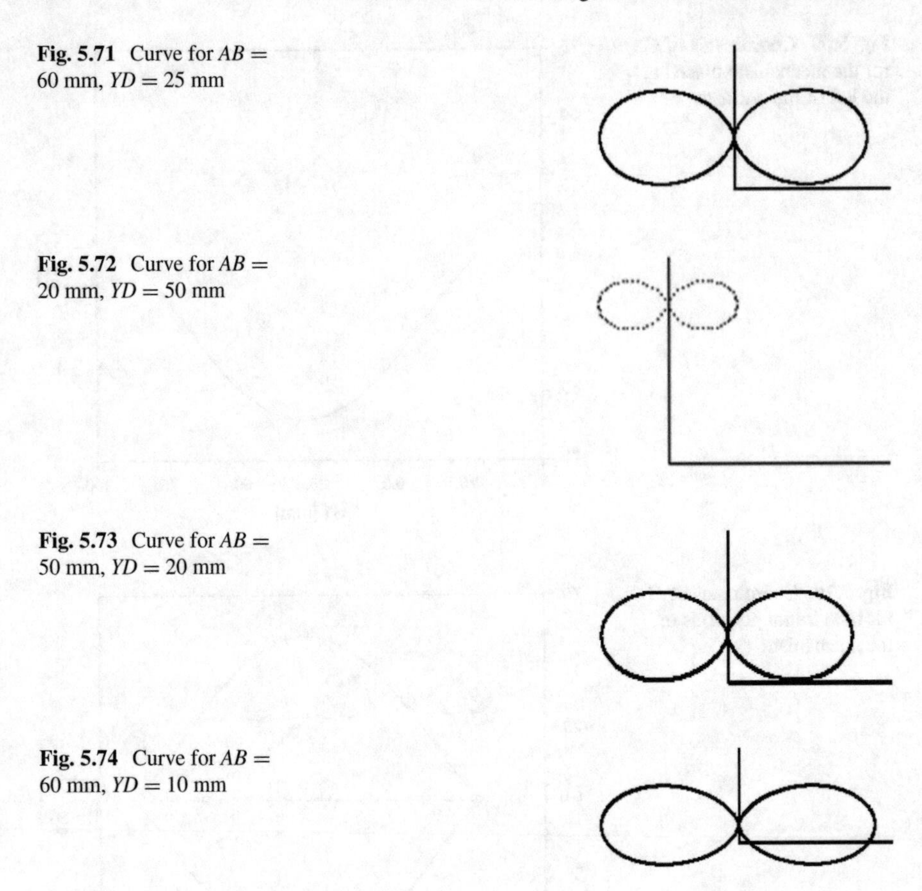

Fig. 5.71 Curve for $AB =$ 60 mm, $YD = 25$ mm

Fig. 5.72 Curve for $AB =$ 20 mm, $YD = 50$ mm

Fig. 5.73 Curve for $AB =$ 50 mm, $YD = 20$ mm

Fig. 5.74 Curve for $AB =$ 60 mm, $YD = 10$ mm

If *YD* decreases more, then the curve intersects the abscissa (Fig. 5.74).
At negative values of *YD*, the curve moves toward $(-y)$ as in Fig. 5.75.
If $YD = 0$, the curve is symmetrical with respect to the abscissa (Fig. 5.76).
For larger values of *YD*, the curve will not close (Fig. 5.77).

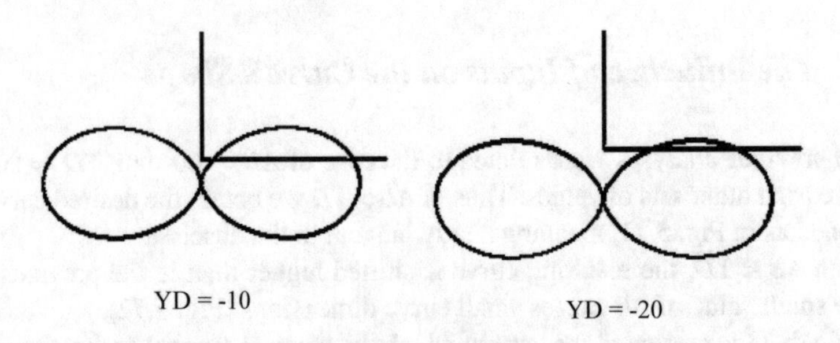

Fig. 5.75 Curve for negative values of *YD*

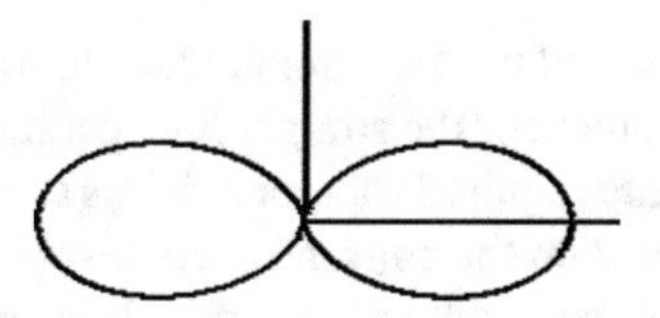

Fig. 5.76 Curve for $AB = 60$ mm, $YD = 0$

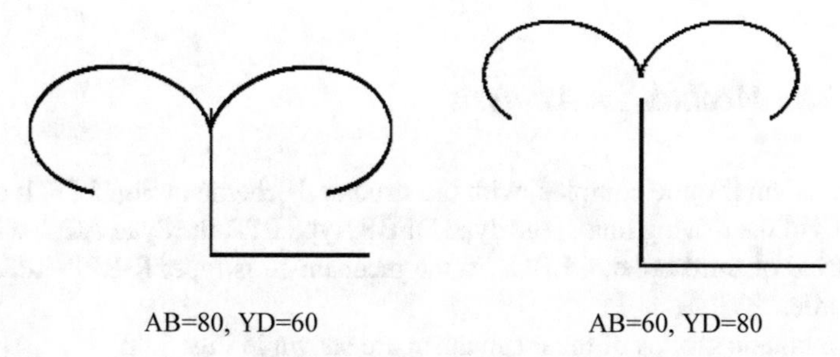

Fig. 5.77 Open curve for larger values of *YD*

5.5 Mechanism for Generating Bernoulli Quartic

5.5.1 The Mechanism Synthesis

Bernoulli quartic was studied in 1687 by Jacques Bernoulli and in 1696 by Leibniz [5].

Geometrically, a line moving with the ends on two concentric circles is always parallel to itself, so that the middle point of this line describes a branch of Bernoulli quartic [11].

The synthesis of the mechanism starts from the following considerations:

- The straight line has variable length, so it takes two sliders on the *C* and *B* (Fig. 5.78).

Fig. 5.78 Mechanism obtained

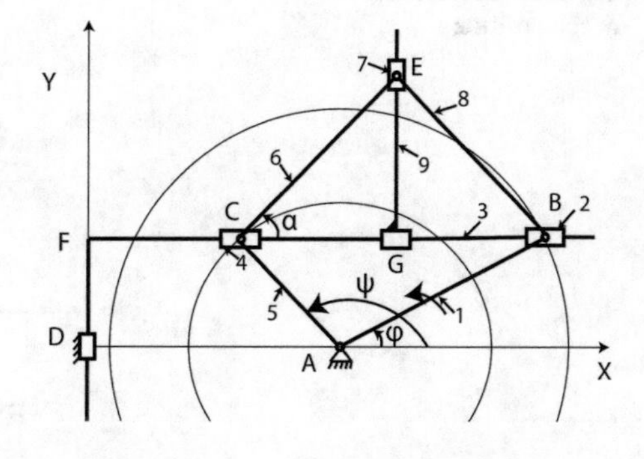

- The straight line has a circular translational motion, so a slider at D is necessary, providing vertical movement of the straight line, parallel to the previous position.
- The sliders at C and B are required rotational joints to ensure rotation of elements AB and AC, with C and B on the concentric circles.
- Point G should be set at the middle of the variable segment BC, so adding BECG kinematic chain.

5.5.2 The Mechanism Analysis

The mechanism is quite complex, with the structural scheme in Fig. 5.79. It consists (Fig. 5.80) of the driving link 1, the dyad DFBB, type RPP, the dyad ACC, type RPR and the triad of third order, CEBG. So the mechanism is type: R-RPP-RPR—triad of third order.

The kinematic groups of the mechanism are shown in Fig. 5.80.

Applying the contours method to the mechanism in Fig. 5.78, Eqs. (5.40)–(5.54) are written:

$$x_B = x_A + AB\cos\varphi \tag{5.40}$$

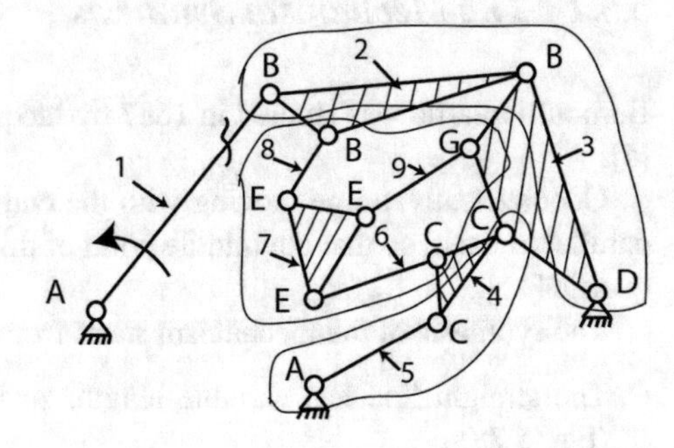

Fig. 5.79 Structure of the mechanism

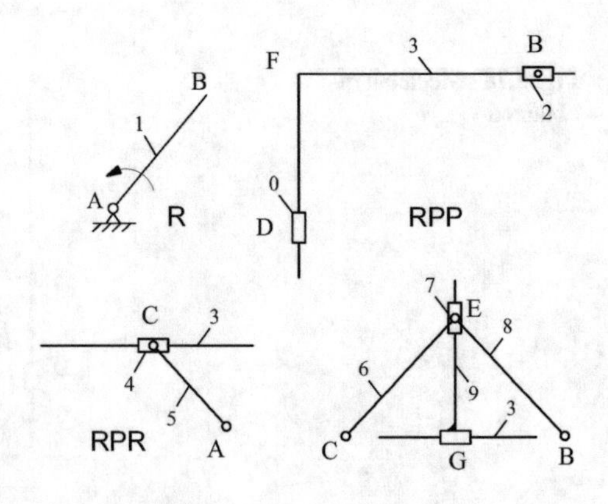

Fig. 5.80 Mechanism decomposition

$$y_B = y_A + AB \sin \varphi \tag{5.41}$$

$$x_D + FB = x_B \tag{5.42}$$

$$y_F = y_B \tag{5.43}$$

$$y_C = y_B = DF = y_A + AC \sin \Psi \tag{5.44}$$

$$x_C = x_A + AC \cos \Psi \tag{5.45}$$

$$CB = x_B - x_C \tag{5.46}$$

$$EB^2 = EC^2 + CB^2 - 2EC \cdot CB \cos \alpha \tag{5.47}$$

$$x_E = x_C + EC \cos \alpha \tag{5.48}$$

$$EG = EC \sin \alpha \tag{5.49}$$

$$CG = EC \cos \alpha \tag{5.50}$$

$$x_G = x_C + CG \tag{5.51}$$

$$y_G = y_B \tag{5.52}$$

$$y_E = y_B + EG \tag{5.53}$$

$$EB = EC \tag{5.54}$$

5.5.3 Results Obtained

In Fig. 5.81, the outcome mechanism (in one position) is shown for the following initial data: $XA = 75$; $AB = 68$; $AC = 42$; $FG = 92$; $BE = 63$; $EC = 65$; $S = -1$; $S1 = 1$ (mm). S and $S1$ are the signs of the square roots that appear to solve the above system.

Because the mechanism moves only in a narrow field, two lengths have changed the values: $AB = 50$, $AC = 130$ (mm), resulting the successive positions of the mechanism in Fig. 5.82.

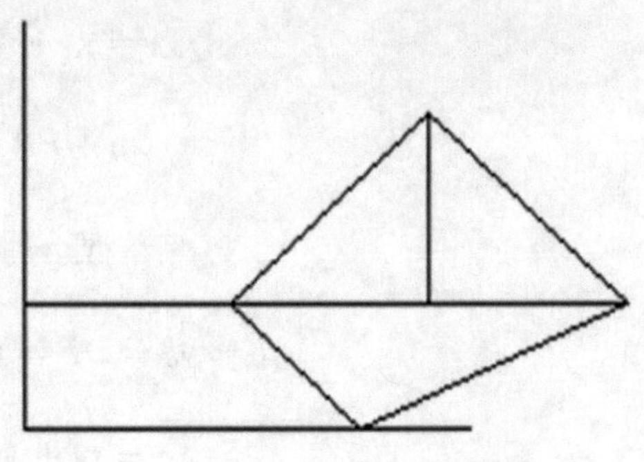

Fig. 5.81 Mechanism in a position

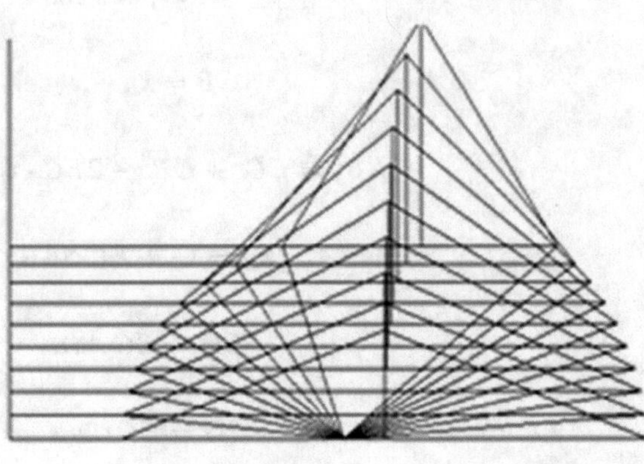

Fig. 5.82 Successive positions

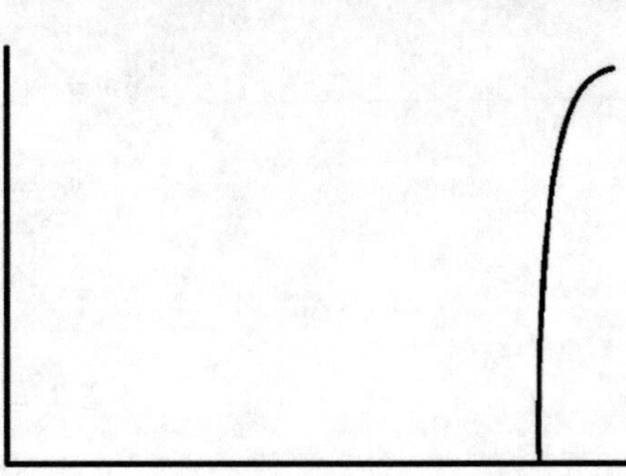

Fig. 5.83 A portion of the curve

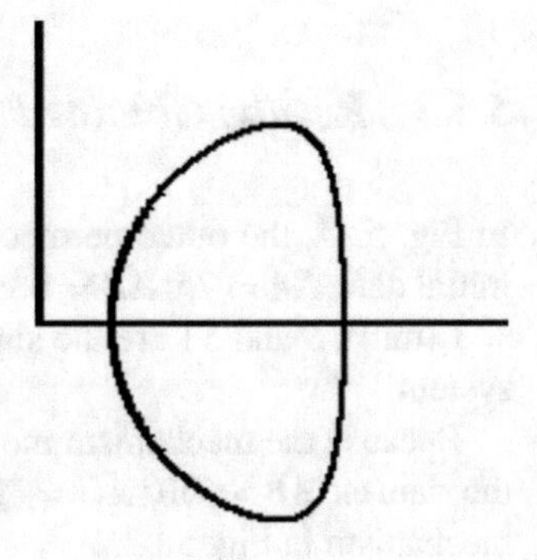

Fig. 5.84 Curve for $AB = 50$ mm, $AC = 69$ mm

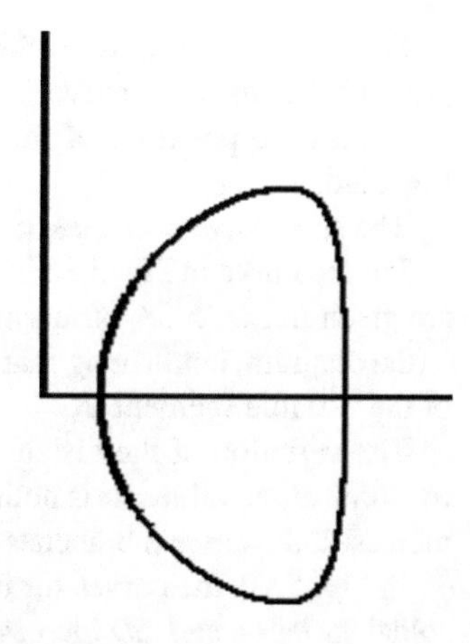

Fig. 5.85 Required curve ($AB = 55$ mm, $AC = 69$ mm)

It appears that the mechanism is well built since G is always in the middle of the variable length BC. However, it also noted the following:

- The obtained curve (Fig. 5.83) is not the required one.
- $AC > AB$ unlike the indication in Fig. 5.78.

By multiple attempts, Fig. 5.84 resulted at $AB = 50$ mm, $AC = 69$ mm, achieving a branch of the curve, similar to the desired curve.

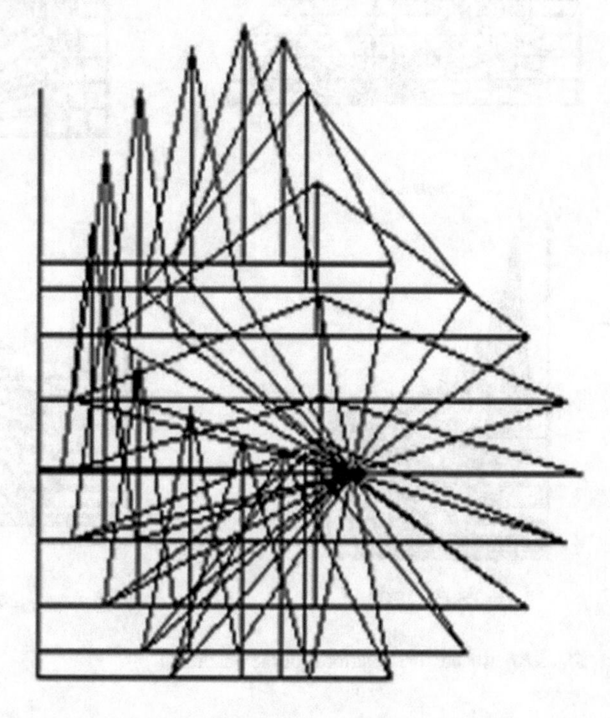

Fig. 5.86 Successive positions for the required curve

For $AB = 55$ mm, $AC = 69$ mm, the curve in Fig. 5.85 was obtained, which is precisely the required curve.

Successive positions of the mechanism for the curve in Fig. 5.85 are given in Fig. 5.86.

The successive positions, detailed on the four quadrants, are shown in Fig. 5.87.

For the curve in Fig. 5.85, the variations of the sliders *CG* and *BG* displacements are given in Fig. 5.88, observing that they are equal, so they are on the same curve in the diagram, confirming that the mechanism ensures that point *G* is in the middle of the variable segment *BC*.

The variation of the distance *CB* is given in Fig. 5.89, similar to Fig. 5.88, but for some other values. It is noticed that *CB* decreases, reaches a minimum, and then increases, the curve's branches being symmetrical.

In Fig. 5.90, the curves for the *EG* and *YG* are observed. The curve *YG* has sinusoidal variation and *ED* increases at the beginning, remains approximately constant in a field, then decreases, its branches being symmetrical.

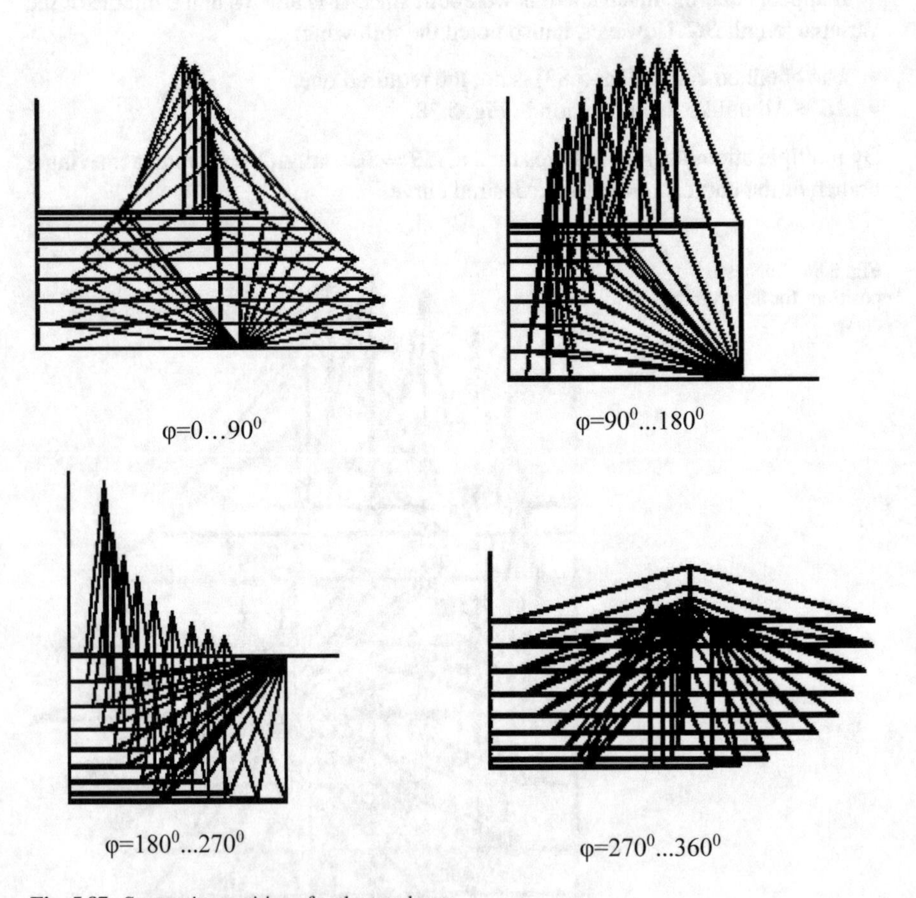

Fig. 5.87 Successive positions for the quadrants

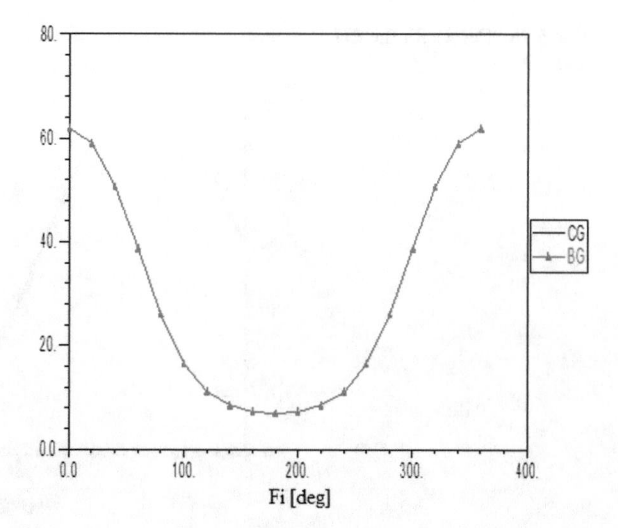

Fig. 5.88 Variations of *CG* şi *BG* (overlapped curves)

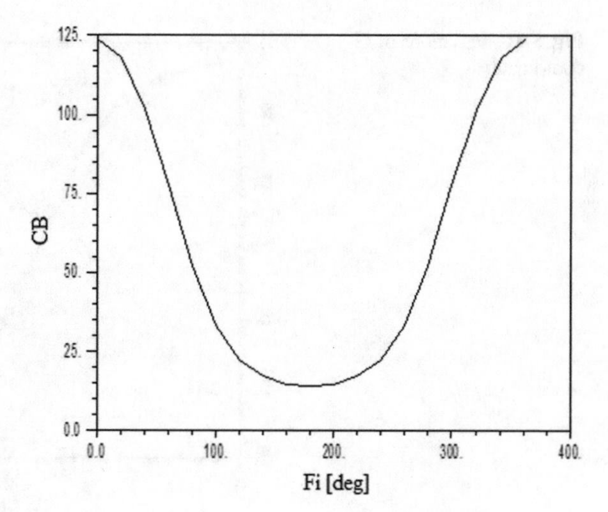

Fig. 5.89 Variation *CB* (φ)

In Fig. 5.91, the variations of the tracer point *G* coordinates are shown. It is noted that *YG* has a sinusoidal variation, while *XG* has a cosinusoidal variation. Curves are continuous, so the mechanism does not break the cycle.

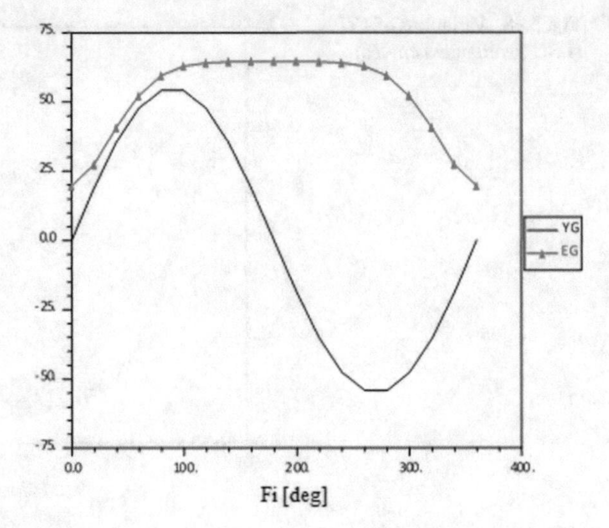

Fig. 5.90 Curves for the *EG* and *YG*

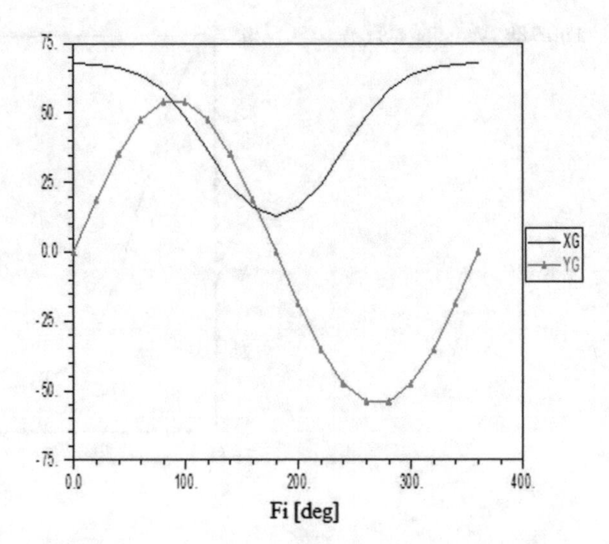

Fig. 5.91 Variations of *G* coordinates

5.6 Mechanism for Generating MacLaurin's Trisectrix

5.6.1 Introduction

Many classic curves obtained by different mathematicians over time are studied in detail. Thus, in [5] the most popular known 2D curves are given, indicating their equations in polar and/or Cartesian coordinates, and some aspects regarding their geometric properties. In nineteenth–twentieth centuries, mechanisms generating such curves as trajectories of some points were designed. Next, an original mechanism that draws Maclaurin's trisectrix is studied, starting from the geometric considerations given in [5].

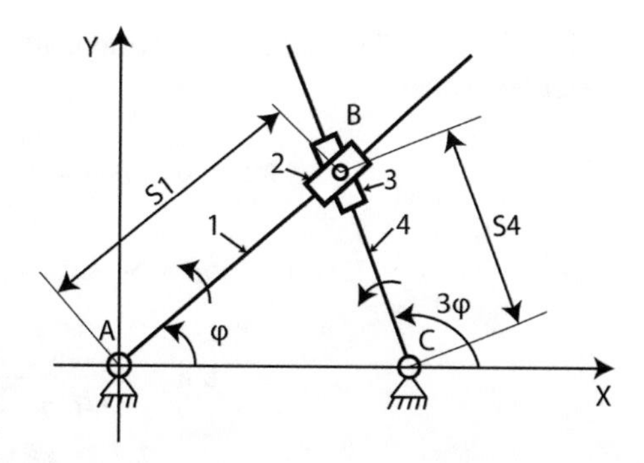

Fig. 5.92 Mechanism obtained

5.6.2 The Mechanism Synthesis

The curve is described by the point of intersection of two lines that each rotates around a fixed point, one with the angular velocity three times higher than the other.

Based on the geometrical construction we built the mechanism in Fig. 5.92 as follows [12]:

- Lines *AB* and *CB* rotate around points *A* and *C*.
- The lengths of the segments are variable, so sliders 2 and 3 were provided.
- The angle between the lines is variable; therefore, between sliders, a rotational joint (at *B*) was provided.
- The straight line *AB* has the angle φ as generalized coordinate and *CB*, respectively, the angle 3 φ.
- The correlation of the two angles can be either through a gear or electromagnetic bases (rotations of the driving motors).

5.6.3 The Mechanism Analysis

Structurally, the mechanism (Fig. 5.93) consists of two driving links *AB* and *CB* and one dyad-type PRP, i.e., R-R-PRP.

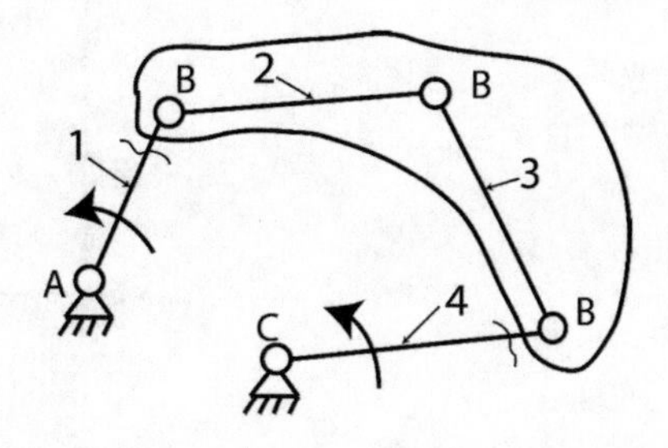

Fig. 5.93 Structure of the mechanism

Equations (5.55)–(5.58) are written, based on the contours method, for the mechanism analysis:

$$x_B = S_1 \cos\varphi = x_C + S_4 \cos 3\varphi \tag{5.55}$$

$$y_B = S_1 \sin\varphi = S_4 \sin 3\varphi \tag{5.56}$$

$$\tan\varphi = \frac{S_4 \sin 3\varphi}{x_C + S_4 \cos 3\varphi} \tag{5.57}$$

$$S_4 = \frac{-x_C \tan\varphi}{\cos 3\varphi \tan\varphi - \sin 3\varphi} \tag{5.58}$$

5.6.4 Results Obtained

From Fig. 5.92 and Eqs. (5.55)–(5.58), it can be noticed that the only constant is the XC dimension. The value $XC = 50$ mm was chosen, and the desired curve (Fig. 5.94) was obtained. In Fig. 5.95, the successive positions of the generating mechanism are shown.

In Fig. 5.96, the displacement variations $S1$ (extreme left, top curve) and $S4$ (denoted by $S2$) are shown. Jumps to infinite for $\varphi = 90°$ and $270°$ are noticed. The program limited the jumps to 300 mm in order to obtain the diagram. Outer branches of the curve tend to be infinite, as shown in Fig. 5.94.

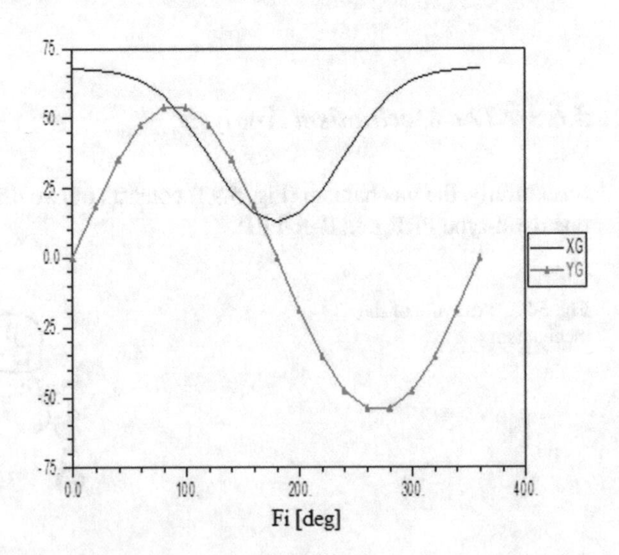

Fig. 5.94 Generated curve

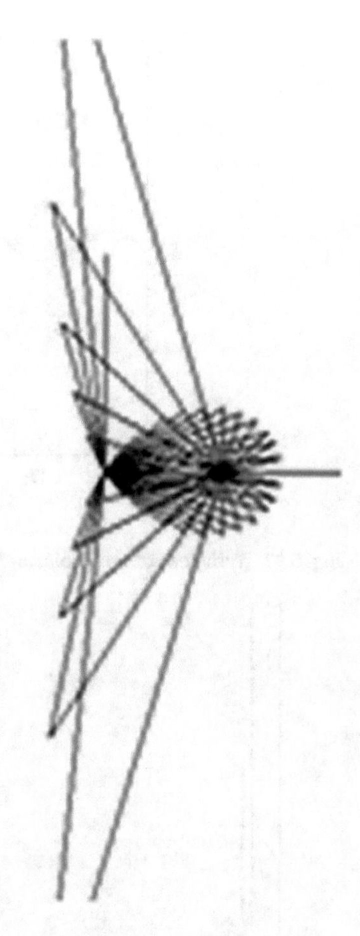

Fig. 5.95 Successive positions

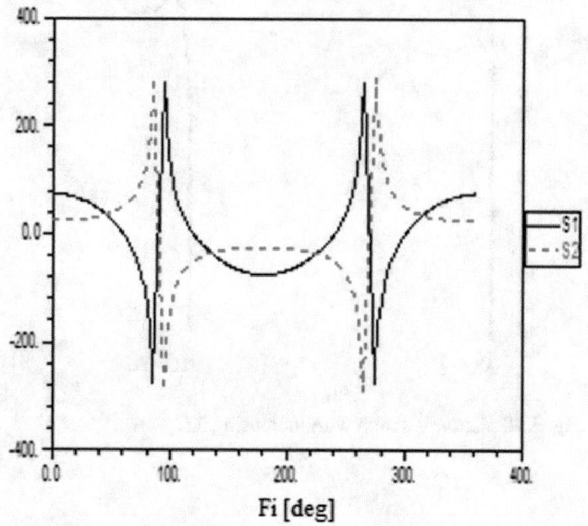

Fig. 5.96 Variations of $S1$ (φ), $S2$ (φ)

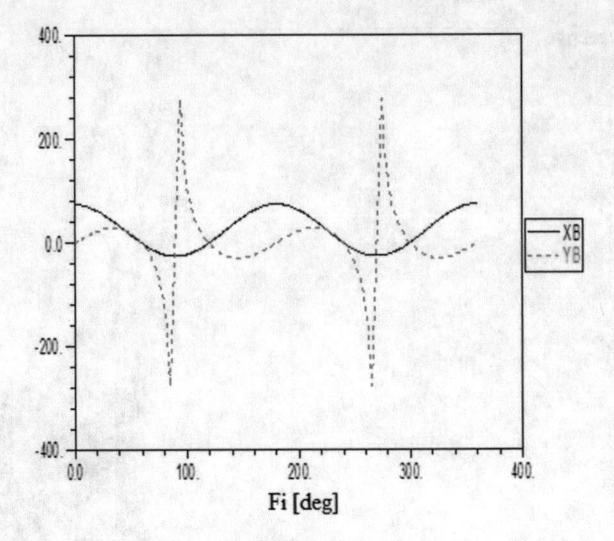

Fig. 5.97 Variations of *B* coordinates for $XC = 50$ mm

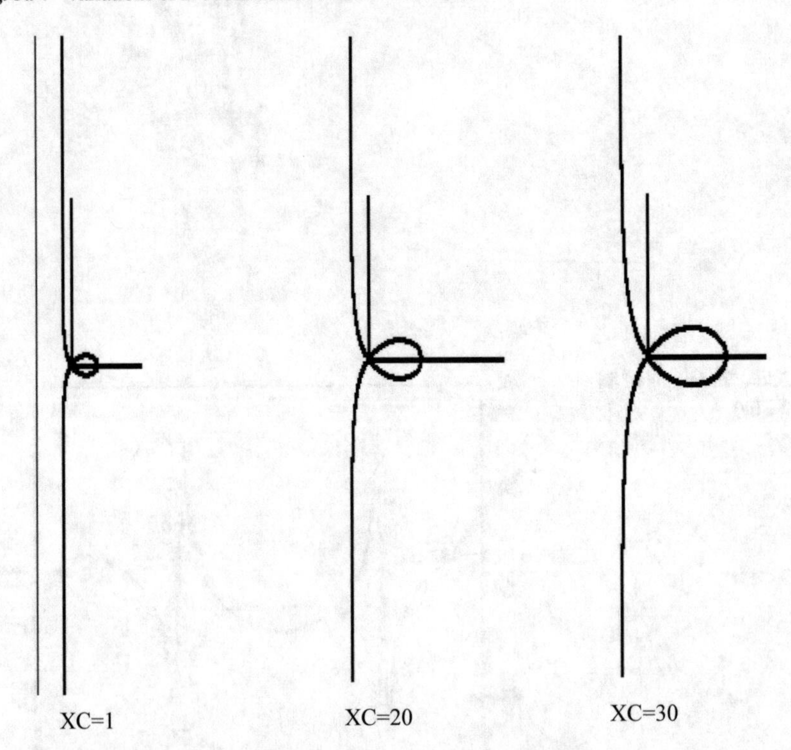

Fig. 5.98 Curve's shape with increasing *XC*

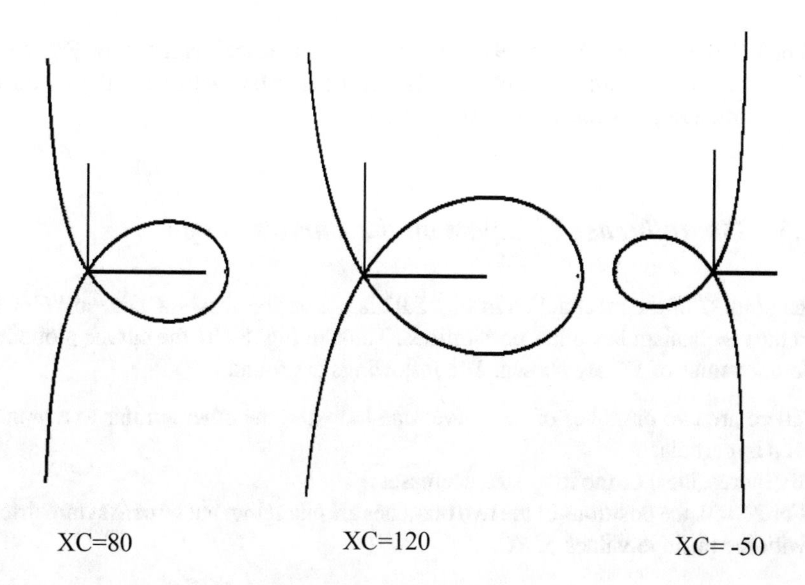

Fig. 5.99 Curve's shape for different values of *XC*

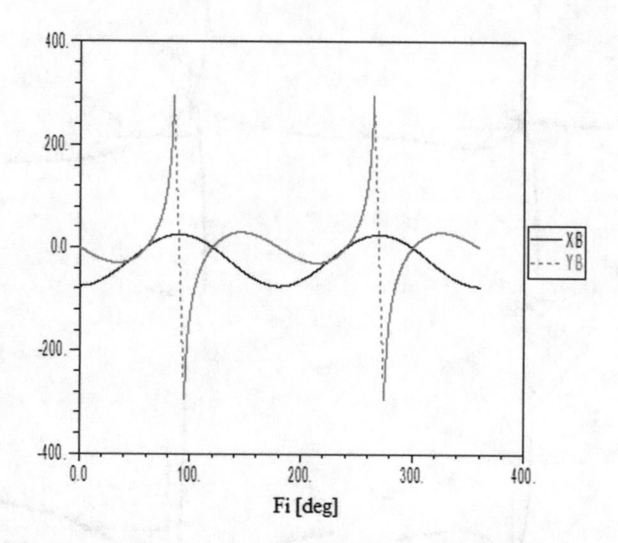

Fig. 5.100 Coordinates of *B* for $XC = -50$

Variations of the coordinates of tracer point (*B*) are given in Fig. 5.97. It is noted that the XB variation is cosinusoidal type and *YB* should be sinusoidal type for $XC = 50$ mm.

If the value of *XC* changes, the curve generated is of the same type, but with different dimensions. In Figs. 5.98 and 5.99, there is an increasing loop size and the distance from the left branch of the curve, proportionally with the increasing of *XC*.

For $XC = -50$ mm (Fig. 5.99), the curve is symmetrical with that of Fig. 5.94. In this case, the variation curve of *XB* (Fig. 5.100) is also negative and the *YB* curve has the same jumps as for $XC = 50$ (Fig. 5.97).

5.6.5 The Influence of Inputs on the Curve's Shape

If the point *C* of the mechanism in Fig. 5.92 is not on the abscissa axis, so $YC \neq 0$, then the mechanism has other possibilities. Thus, in Fig. 5.101 the curves plotted at different values of *YC* are shown. The followings are found:

- There are two branches of the curve: one loop and the other similar to a branch of a hyperbola;
- By increasing *YC*, the loop size decreases;
- For $YC < 0$, the positions of the two branches are changing, but remain symmetrical with the positive values of *YC*.

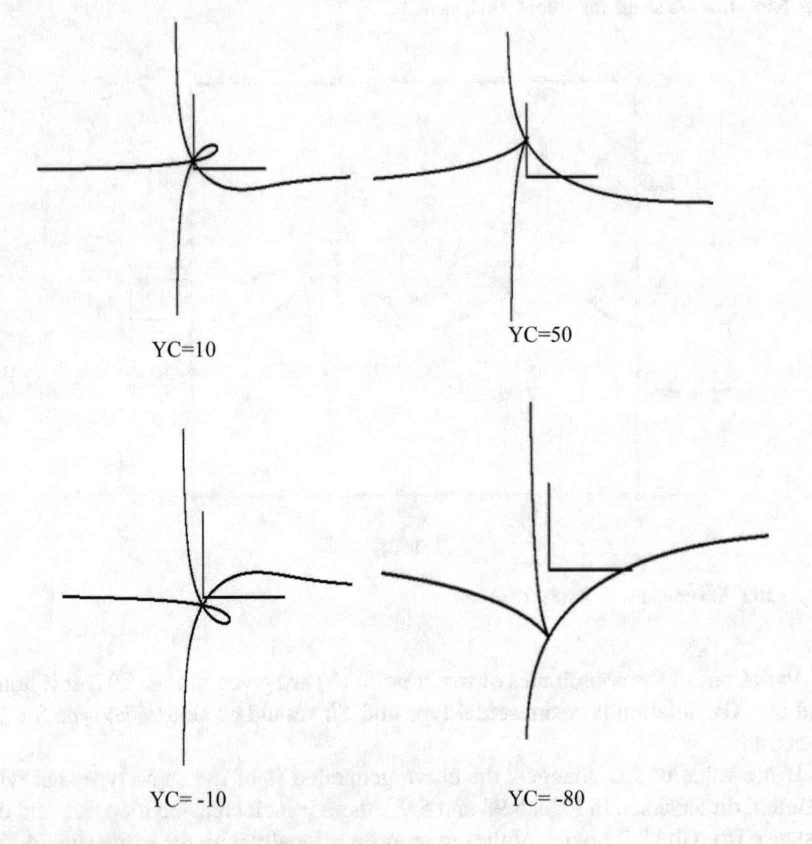

Fig. 5.101 Influence of YC upon the curve's shape

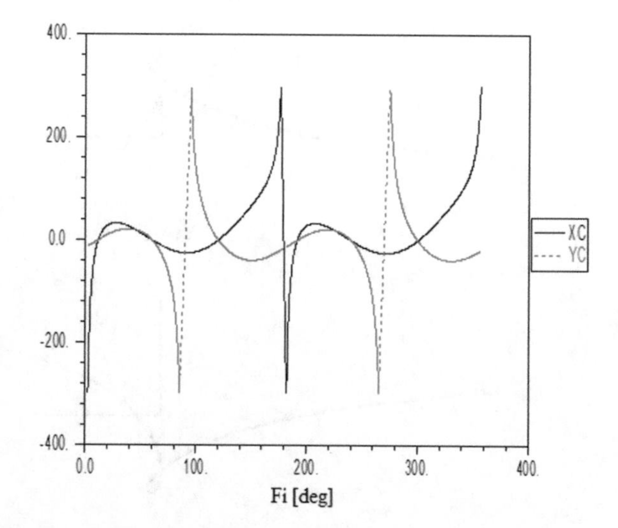

Fig. 5.102 Coordinates of C

In Fig. 5.102, the coordinate variations of tracer point for the $YC = 10$ mm are shown. This time, both XC and YC are areas that tend to infinite. Thus, these trends occur for $\varphi = 0°$, $180°$, $360°$ in case of XC, and for $\varphi = 90°$, $270°$, in case of YC. In Fig. 5.103, the mechanism positions for $YC = -80$ mm are represented. Areas which increase to infinite are noticed.

In Fig. 5.104, curves obtained for 2φ (not 3φ), at different values of YC, are shown.

In Fig. 5.105, the curves of variation of the tracer point coordinates B (for $YC = -50$ mm) are shown, observing that it increases to infinite just for XB curve, which has a break around $\varphi = 180°$, and jumps to infinite at $\varphi = 0°$ and $360°$. These findings can be seen in the successive positions of the mechanism of Fig. 5.106, where it is noted the area where the mechanism does not work.

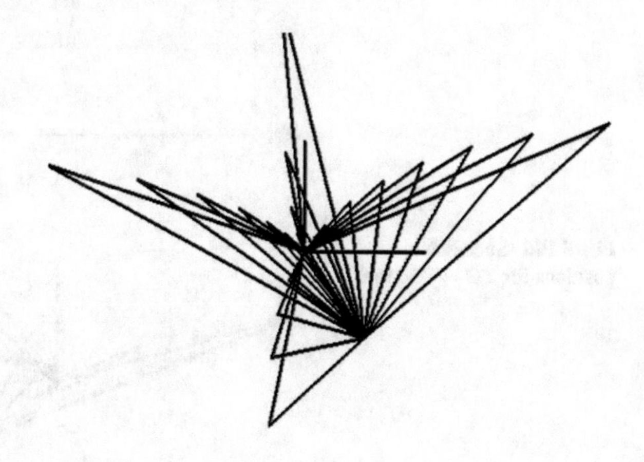

Fig. 5.103 Successive positions

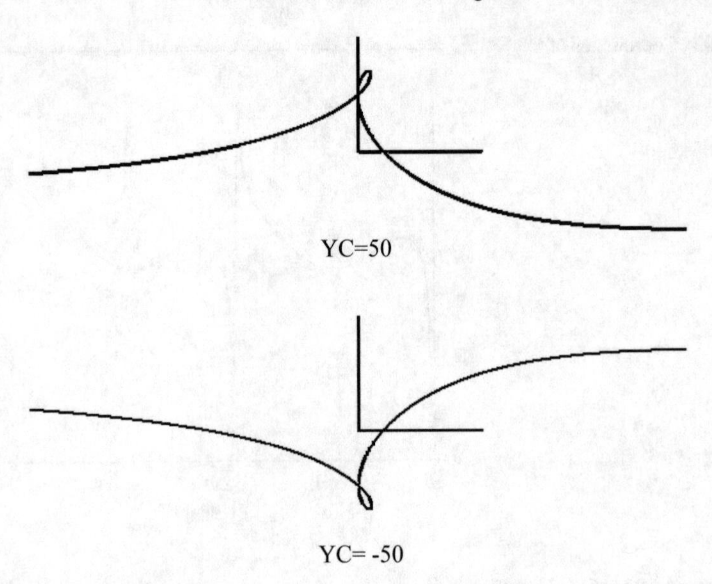

Fig. 5.104 Curve for $YC = -50$

Fig. 5.105 Variation of the tracer point coordinates B for $YC = -50$ mm

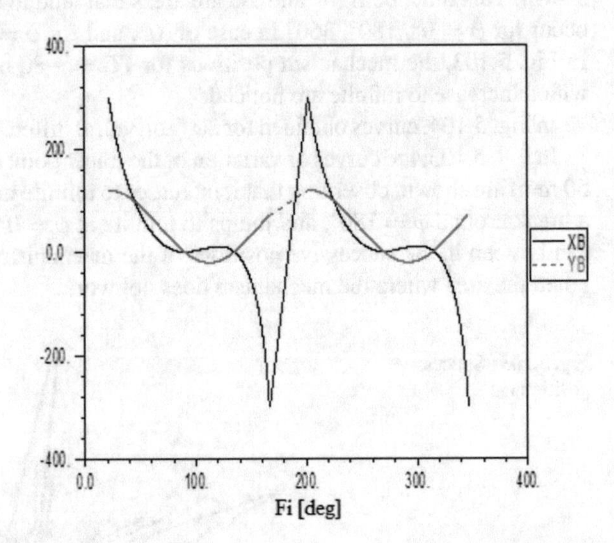

Fig. 5.106 Successive positions for YC = −50 mm

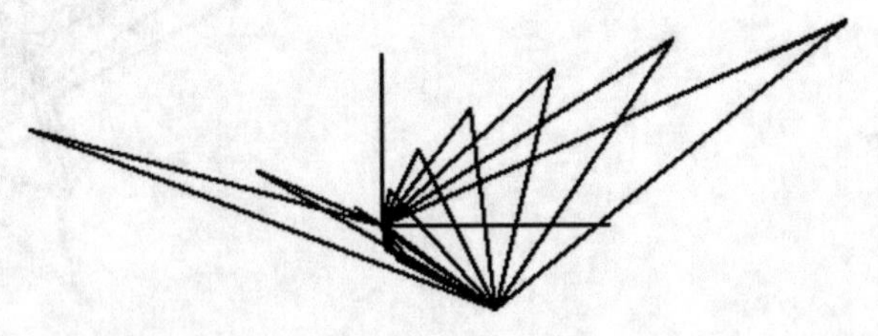

Fig. 5.107 Mechanism obtained

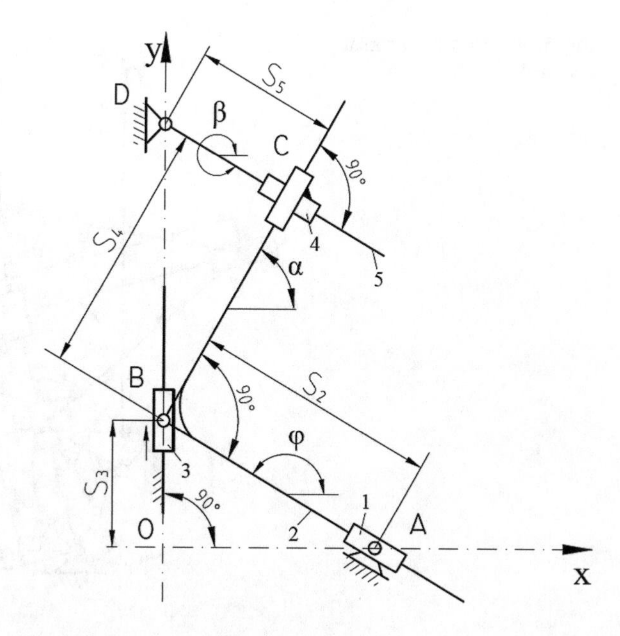

5.7 Mechanism for Generating the Ophiuride

5.7.1 Introduction

The ophiuroid is a curve which has been studied in 1808 by Uhlhorn [5]. In [5], the geometrical considerations for generating the ophiuroid are shown. Two circles with different radii are considered, the inner circle being tangent to the exterior one at a point through which a straight line rests on the outer circle and on the fixed line which is parallel to the ordinate.

In [4], an original mechanism (Fig. 5.107) which generates the ophiuroid (point *C*) is presented, starting from a locus problem.

5.7.2 The Mechanism Synthesis

The above details are concise, so that the mechanism synthesis was difficult. We start from the *AC* outer circle (Fig. 5.108) and the *CD* straight line was drawn, which rests with *B* point on the circle and with *D* point on the fixed straight line which is parallel to the ordinate. For *B* to be always on the circle, the *AB* radius was drawn and in *B* a slide was placed, so that point *B* will translate on the *CD* straight line. Also, in *D* two slides were placed, one allowing the variation of *CD* length and the other providing the movement of *D* on the fixed straight line.

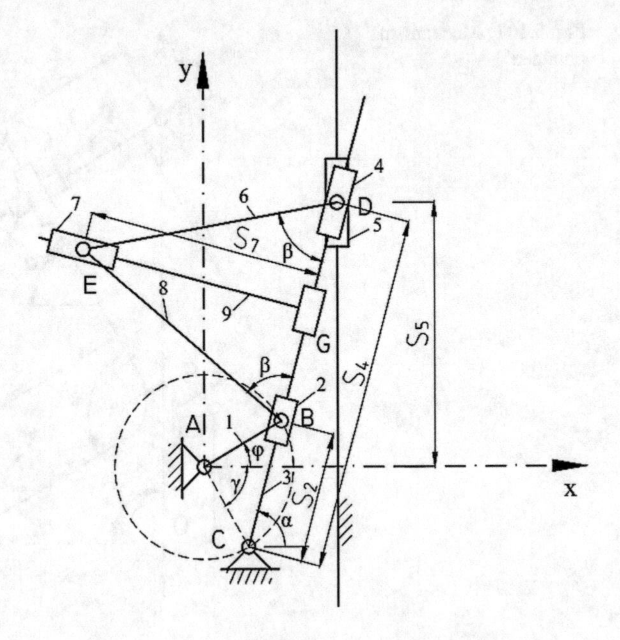

Fig. 5.108 New mechanism obtained

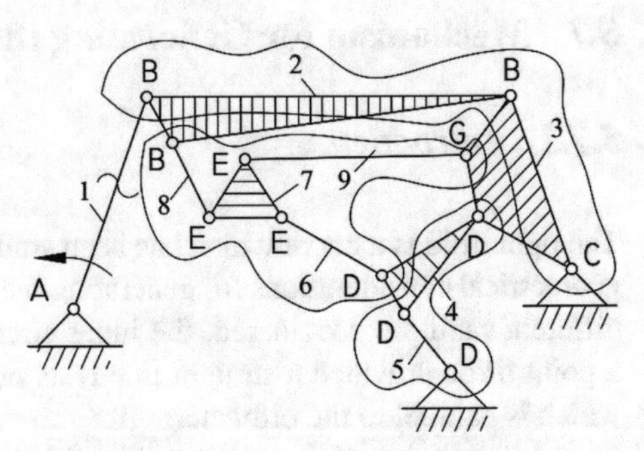

Fig. 5.109 Structural diagram

With no geometric details, the *G* tracing point was considered in various positions, by multiple attempts, observing that it must be at the midpoint of the *BD* segment with variable length. Thus, the additional GEBD kinematic chain was constructed.

The position *C* was adopted on the circle (by calculation, satisfying the equation of the circle), and by experiments, the fixed line position (vertical through *D*) was set.

5.7.3 The Mechanism Analysis

Structurally, the mechanism is quite complicated. From the structural diagram (Fig. 5.109) and from the decomposition in kinematic groups (Fig. 5.110), it appears that the mechanism is R-RPR-PRP—triad of third order.

The angle of $AC = AB$ was adopted with the x-axis (counter-clockwise), which is 300°, resulting in the coordinates of C.

Equations (5.59)–(5.62) were written for the mechanism analysis, according to the contours method:

$$\gamma = \tan^{-1}\frac{y_A - y_C}{x_C - x_A} \tag{5.59}$$

$$S_2^2 = AB^2 + AB^2 - 2AB \cdot AB\cos(\gamma + \varphi) \tag{5.60}$$

$$\cos\alpha = \frac{x_B - x_C}{S_2} \tag{5.61}$$

$$\sin\alpha = \frac{y_B - y_C}{S_2} \tag{5.62}$$

Avoiding the function $\tan^{-1}$, which changes its values depending on the trigonometric quadrants, the coordinates of B were calculated with Eqs. (5.63), (5.64):

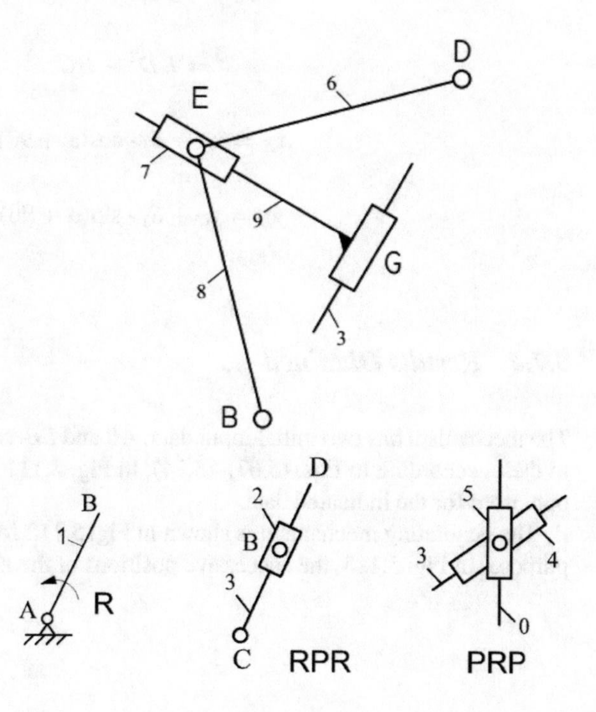

Fig. 5.110 Mechanism decomposition

$$x_B = AB \cos \varphi \quad (5.63)$$

$$y_B = AB \sin \varphi \quad (5.64)$$

Then $S4$ and $S5$ were calculated with Eqs. (5.65), (5.66):

$$x_D = x_C + S_4 \cdot \cos \alpha = \text{const.} \quad (5.65)$$

$$y_D = y_C + S_4 \cdot \sin \alpha = y_A + S_5 \quad (5.66)$$

Further on, the triad position was calculated with Eqs. (5.67)–(5.74):

$$ED = EB \quad (5.67)$$

$$ED^2 = EB^2 + (S_4 - S_2)^2 - 2 \cdot EB \cdot (S_4 - S_2) \cos \beta \quad (5.68)$$

$$BG = EB \cdot \cos \beta \quad (5.69)$$

$$x_G = x_B + BG \cdot \cos \alpha \quad (5.70)$$

$$y_G = y_B + BG \cdot \sin \alpha \quad (5.71)$$

$$S_7^2 = ED^2 - BG^2 \quad (5.72)$$

$$x_E = x_G + S_7 \cdot \cos(\alpha + 90) \quad (5.73)$$

$$y_E = y_G + S_7 \cdot \sin(\alpha + 90) \quad (5.74)$$

5.7.4 Results Obtained

The mechanism has two initial input data: *AB* and *EB*, other results being dependent to these, according to Eqs. (5.67)–(5.74). In Fig. 5.111, it is presented the resulting ophiuroid for the indicated data.

The generating mechanism is shown in Fig. 5.112 for a position, so it fits for the purpose. In Fig. 5.113, the successive positions of the mechanism are given.

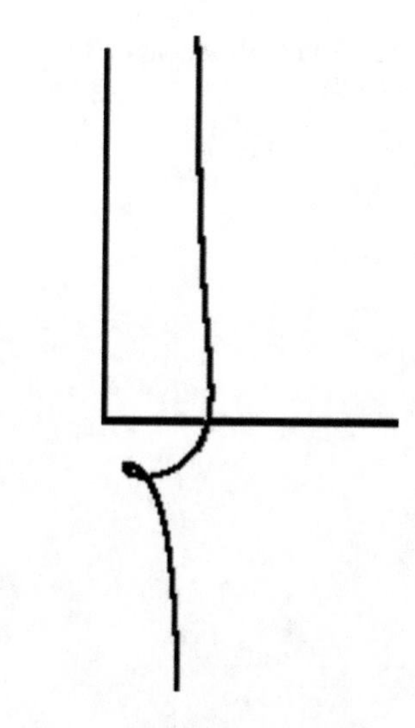

Fig. 5.111 Curve for $AB =$ 35 mm, $ED = 180$ mm

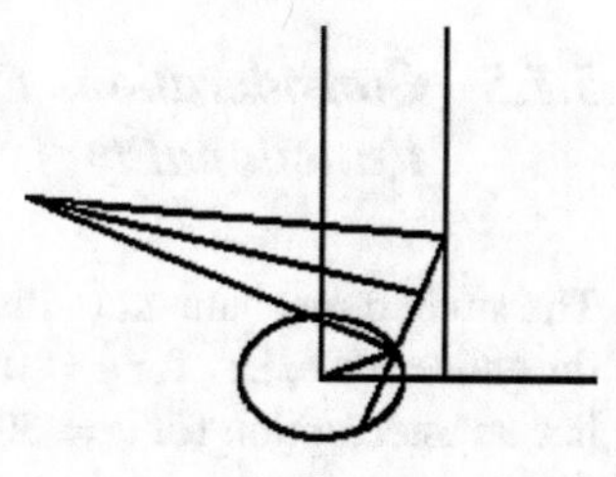

Fig. 5.112 Mechanism in a position

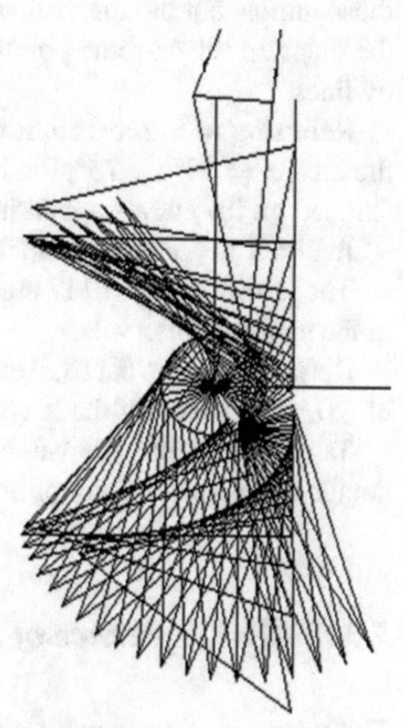

Fig. 5.113 Successive positions of the mechanism

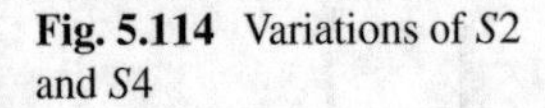

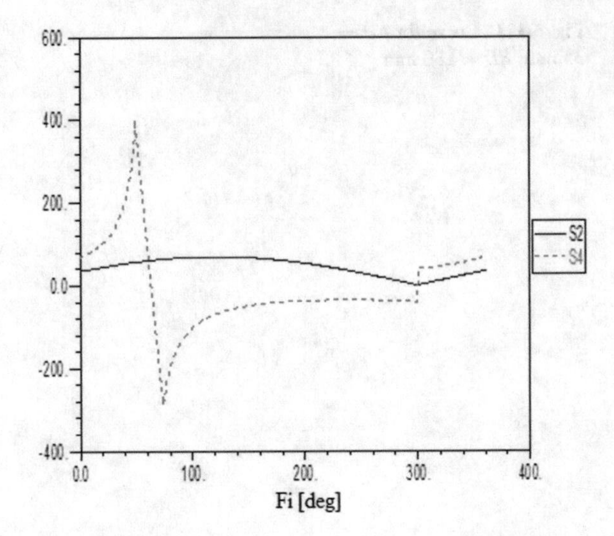

Fig. 5.114 Variations of *S*2 and *S*4

5.7.5 *Considerations Regarding the Mechanism Functionality*

The studied mechanism (with *AB* = 35 mm and *ED* = 180 mm) does not work for the entire cycle, i.e., for $\varphi = 0°$ … 360°. Thus, from Fig. 5.114, it is observed that *S*4 has an interruption for $\varphi = 50°$ … 75°and at $\varphi = 300°$. The *S*2 curve does not have these jumps, but the mechanism does not work for the specified intervals, so that on the diagram the extreme points of the area in which it does not operate were joined by lines.

Referring to Fig. 5.115, it is found that the curves for *S*5 and *S*7 have jumps in the area $\varphi = 50°$ … 75°, the higher jump being for *S*5 at 300 mm. This value was skipped on the curves appearing one small segment of line.

In Fig. 5.116, a large jump for *YG* curve in the same area ($\varphi = 50°$ … 75°) appears.

The jumps in Fig. 5.117 are very clear, where both *XE* and *YE* do not have values in the same subintervals.

Referring to Fig. 5.118, it can be observed that the values of *BG* are half the values of *BD*, so *G* is at half the length of the *BD* variable.

Table 5.1 reveals the values of variables on the mechanism, observing in more details the mentioned jumps and also the irregular variations of these characteristics.

5.7.6 *The Influence of Input Data on the Curve's Shape*

The influence of the input data on the generated curve was studied. Thus, in Fig. 5.119, a small curve resulting from the low values of *AB* and *ED* is shown. If *AB* remains

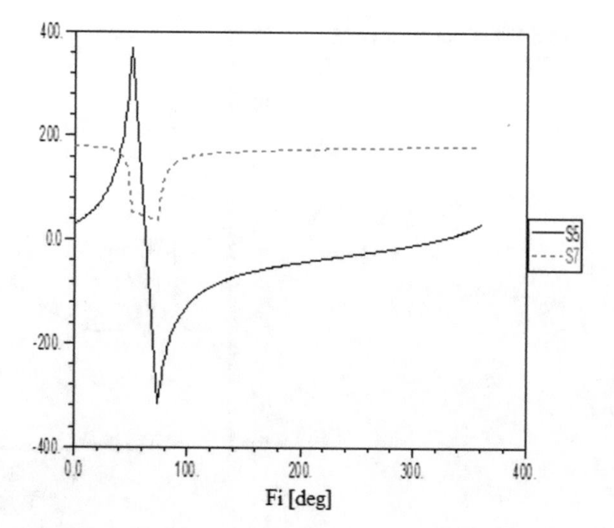

Fig. 5.115 Variations of *S*5 and *S*7

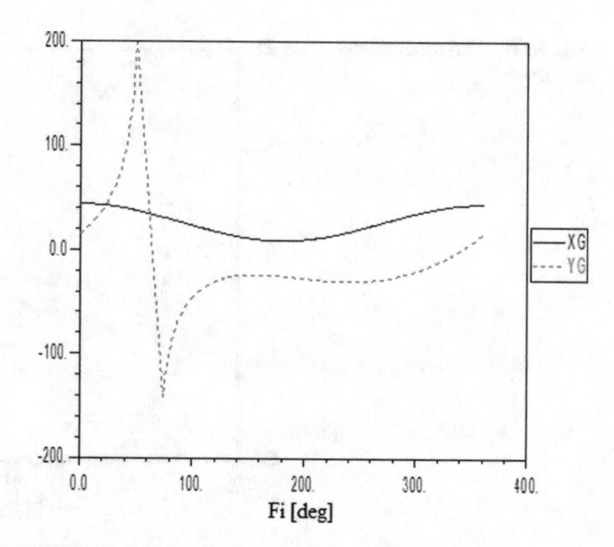

Fig. 5.116 Coordinates of *G*

constant = 20 mm and *ED* increases, the ophiuroid has the same loop, but increases its branches that have as asymptote the fixed line passing through *D* (Fig. 5.119).

If *AB* is increased, i.e., *AB* = 30 mm the loop moves more under the abscissa, and the branches are increasing with the increasing of *ED* (Fig. 5.120).

If *AB* increases, and *ED* is maintained constant but at higher values, i.e., *ED* = 200 mm the loop is moving below, but it is increasing. However, the lower branch of the ophiuroid is reduced (Fig. 5.121).

At values too high for *AB*, with *ED* = 200 mm = const., the lower branch of the curve is reduced, leaving the loop incomplete (Fig. 5.122).

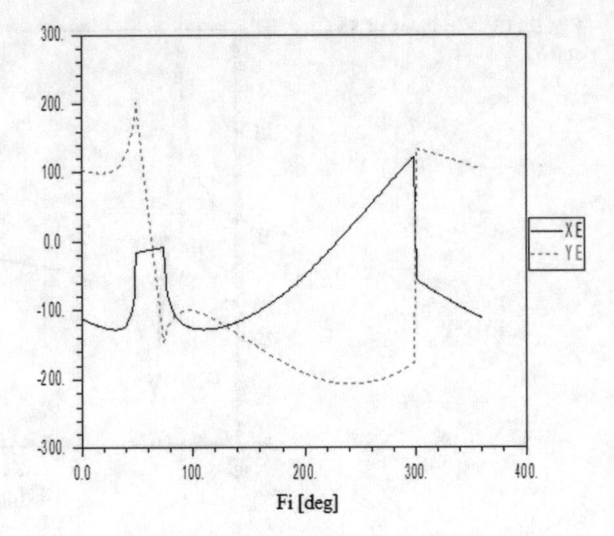

Fig. 5.117 Coordinates of E

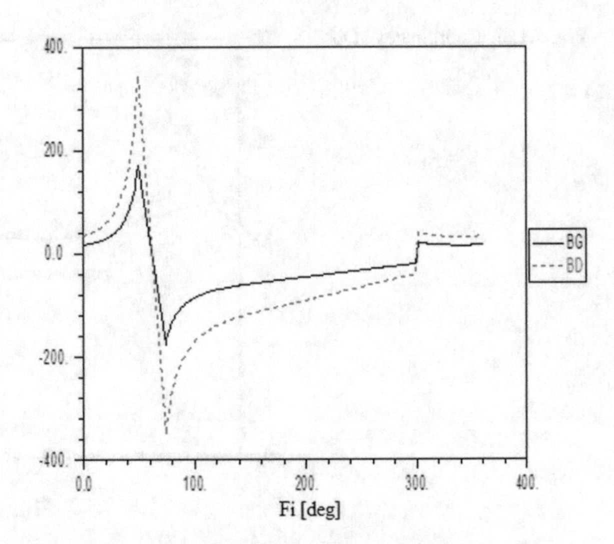

Fig. 5.118 Variations of BG and BD

For $AB < 0$, the curve becomes symmetrical to the one with $AB > 0$, with the same value, but with different positioning (Fig. 5.123a). By comparing it to Fig. 5.121 ($AB = 50$), it is noted the curve is symmetrical in shape, but it is positioned from the fourth quadrant to the IInd quadrant, maintaining the distances from the abscissa and ordinate, but with opposite signs. If $AB = 50$ mm and $ED = -200$ mm, Fig. 5.123b is obtained, which is identical to Fig. 5.121 for $AB = 50$ mm. This means that the triad position in BD's left or right side does not influence the ophiuroid shape.

Table 5.1 *BG* and *BD* displacements

F_i	S2	S4	S5	S7	BG	BD
0	35.00012	69.9993	30.31014	179.1473	17.49959	34.99918
5	37.61109	75.79801	36.92262	178.9845	19.09346	38.18693
10	40.15046	82.81613	44.74587	178.7314	21.33284	42.66567
15	42.61341	91.45833	54.18545	178.3355	24.42246	48.84493
20	44.99524	102.3319	65.84949	177.7024	28.66833	57.33666
25	47.29141	116.3913	80.69327	176.6531	34.54992	69.09984
30	49.49756	135.2276	100.3088	174.8216	42.86503	85.73004
35	51.6095	161.7052	127.5611	171.376	55.04787	110.0957
40	53.62319	201.5531	168.18	164.1011	73.96498	147.9299
45	55.53481	268.1391	235.5341	145.2579	106.3022	212.6043
50	57.34071	401.5668	369.7277	52.69812	172.1131	344.2261
75	64.67161	−268.1486	−296.1655	68.61262	−166.41	−332.8202
80	65.77852	−201.5584	−228.8074	120.5518	−133.669	−267.3369
85	66.76022	−161.7087	−188.1866	139.1061	−114.234	−228.4689
90	67.61483	−135.23	−160.9332	148.7061	−101.422	−202.8448
95	68.34075	−116.393	−141.317	154.4939	−92.3669	−184.7338
100	68.93656	−102.3332	−126.4728	158.3246	−85.6349	−171.2698
105	69.40116	−91.45939	−114.8085	161.031	−80.4303	−160.8605
110	69.73364	−82.81697	−105.3687	163.0401	−76.2753	−152.5506
115	69.93338	−75.7987	−97.54532	164.592	−72.866	−145.7321
120	70	−69.99986	−90.93271	165.8313	−69.99993	−139.9999
125	69.93338	−65.14041	−85.24991	166.8496	−67.5369	−135.0738
130	69.73363	−61.02047	−80.2961	167.7076	−65.3771	−130.7541
135	69.40113	−57.49362	−75.92381	168.4471	−63.4474	−126.8948
140	68.93653	−54.45016	−72.0223	169.0974	−61.6934	−123.3867
145	68.34071	−51.80637	−68.50671	169.6796	−60.0735	−120.1471
150	67.6148	−49.49729	−65.31093	170.2093	−58.556	−117.1121
155	66.76017	−47.47177	−62.38253	170.6979	−57.116	−114.2319
160	65.77846	−45.68906	−59.67942	171.1542	−55.7338	−111.4675
165	64.67155	−44.11634	−57.16739	171.5847	−54.394	−108.7879
170	63.44152	−42.72691	−54.81822	171.9944	−53.0842	−106.1684
175	62.09074	−41.49892	−52.60842	172.3871	−51.7948	−103.5896
180	60.62176	−40.41432	−50.51823	172.7655	−50.518	−101.0361
185	59.03738	−39.45817	−48.53083	173.1319	−49.2478	−98.49554
190	57.34061	−38.61802	−46.63175	173.4877	−47.9793	−95.95863
195	55.5347	−37.88352	−44.80847	173.834	−46.7091	−93.41821
200	53.62308	−37.24601	−43.04996	174.1715	−45.4346	−90.86909

(continued)

Table 5.1 (continued)

F_i	S2	S4	S5	S7	BG	BD
205	51.60938	−36.6983	−41.34646	174.5005	−44.1538	−88.30768
210	49.49745	−36.23446	−39.68924	174.8214	−42.866	−85.7319
215	47.29129	−35.84957	−38.07034	175.134	−41.5704	−83.14086
220	44.99511	−35.53971	−36.48248	175.4381	−40.2674	−80.53482
225	42.61328	−35.30179	−34.91889	175.7336	−38.9575	−77.91507
230	40.15033	−35.13346	−33.37316	176.0201	−37.6419	−75.28378
235	37.61095	−35.03311	−31.8392	176.2972	−36.322	−72.64406
240	34.99998	−34.99976	−30.31108	176.5645	−34.9999	−69.99974
245	32.32238	−35.0331	−28.78296	176.8214	−33.6777	−67.35547
250	29.58325	−35.13344	−27.249	177.0676	−32.3583	−64.71668
255	26.78781	−35.30175	−25.70328	177.3026	−31.0448	−62.08956
260	23.94138	−35.53966	−24.13968	177.5261	−29.7405	−59.48105
265	21.04938	−35.84951	−22.55183	177.7375	−28.4494	−56.89889
270	18.11732	−36.23437	−20.93294	177.9367	−27.1759	−54.35169
275	15.15077	−36.69821	−19.27573	178.1233	−25.9245	−51.84898
280	12.15537	−37.24591	−17.57223	178.2972	−24.7006	−49.40127
285	9.136818	−37.88336	−15.81374	178.4581	−23.5101	−47.02017
290	6.100898	−38.61787	−13.99045	178.6059	−22.3594	−44.71877
295	3.053359	−39.45804	−12.09142	178.7406	−21.2557	−42.5114
305	3.053319	41.4982	−8.013834	178.9707	19.22244	38.44489
310	6.100878	42.72652	−5.804087	179.066	18.31282	36.62565
315	9.136818	44.11611	−3.454928	179.1483	17.48964	34.97929
320	12.15537	45.68878	−0.9429073	179.2174	16.76671	33.53341
325	15.15076	47.47146	1.760134	179.2731	16.16035	32.3207
330	18.11732	49.49695	4.688505	179.3149	15.68981	31.37963
335	21.04939	51.80597	7.884257	179.3419	15.37829	30.75658
340	23.9414	54.44971	11.39978	179.3525	15.25416	30.50831
345	26.78783	57.49309	15.30121	179.3441	15.35263	30.70526
350	29.58326	61.01986	19.67342	179.3124	15.7183	31.43659
355	32.32239	65.13971	24.62716	179.2505	16.40866	32.81732
360	34.99998	69.99902	30.30981	179.1473	17.49952	34.99904

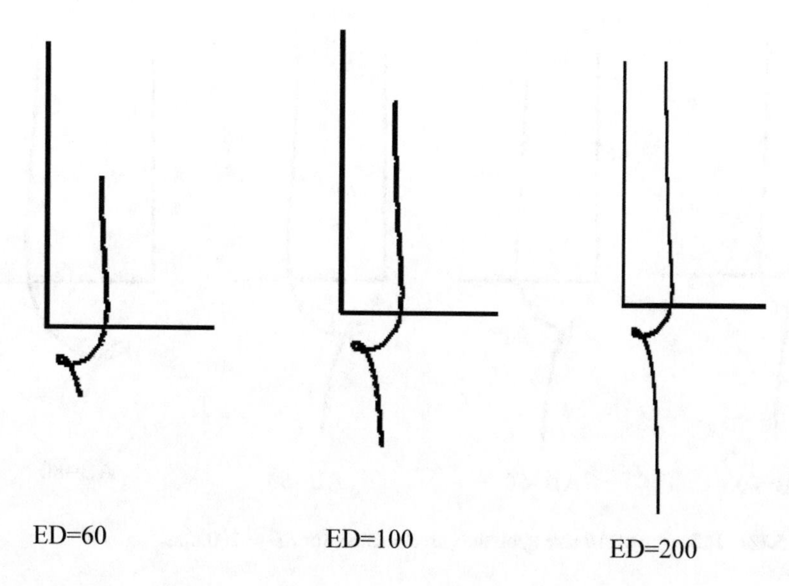

Fig. 5.119 Influence of *ED* size upon the curve's shape, for $AB = 20$ mm

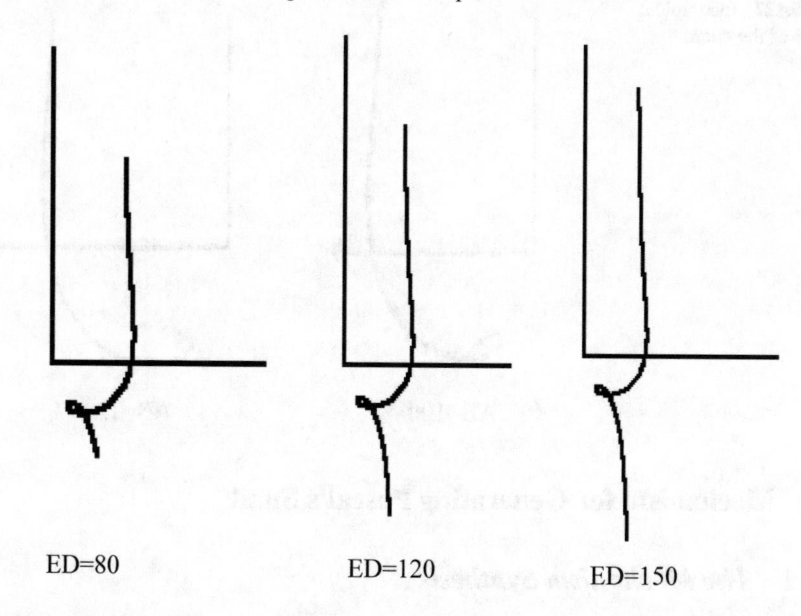

Fig. 5.120 The influence of *ED* size upon the curve's shape, for $AB = 30$ mm

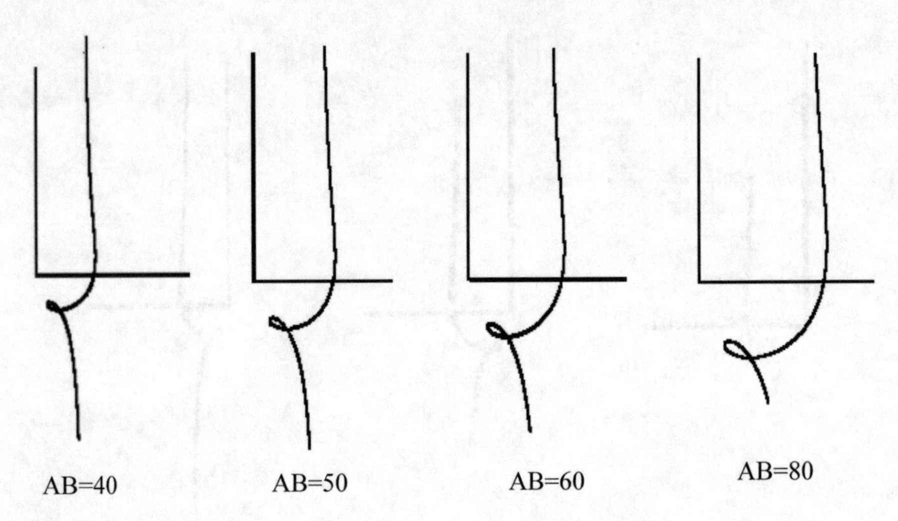

Fig. 5.121 Influence of *AB* size upon the curve's shape, for $ED = 200$ mm

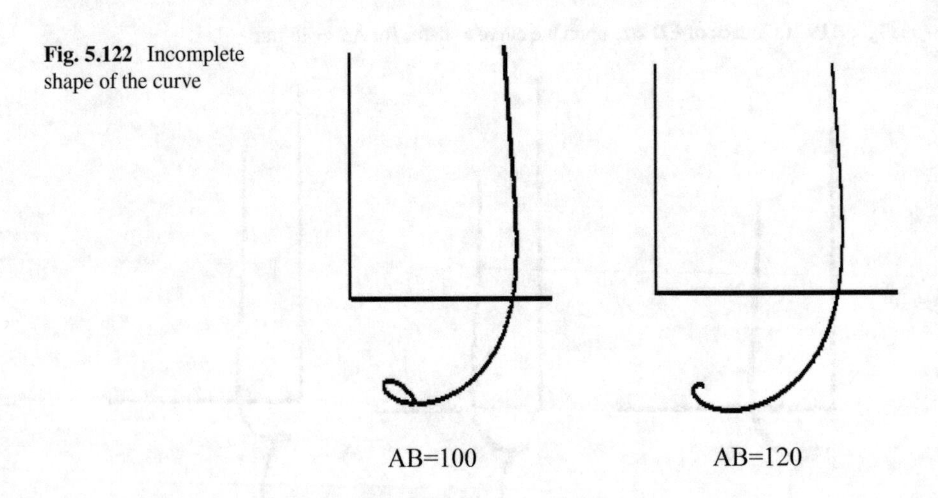

Fig. 5.122 Incomplete shape of the curve

5.8 Mechanism for Generating Pascal's Snail

5.8.1 The Mechanism Synthesis

Studying the mechanisms generating pedals of curves, we reached the mechanism in Fig. 5.124 [13]. The circle with the center at *A* and with radius *AD* has the *EBF* tangent welded to radius *AD*. The perpendicular from *C* is set on the *EF* tangent, which is parallel to *AD*, and an element is mounted in *B*, which is connected with two sliders of length zero. The locus of *B* follows a Pascal's snail [3].

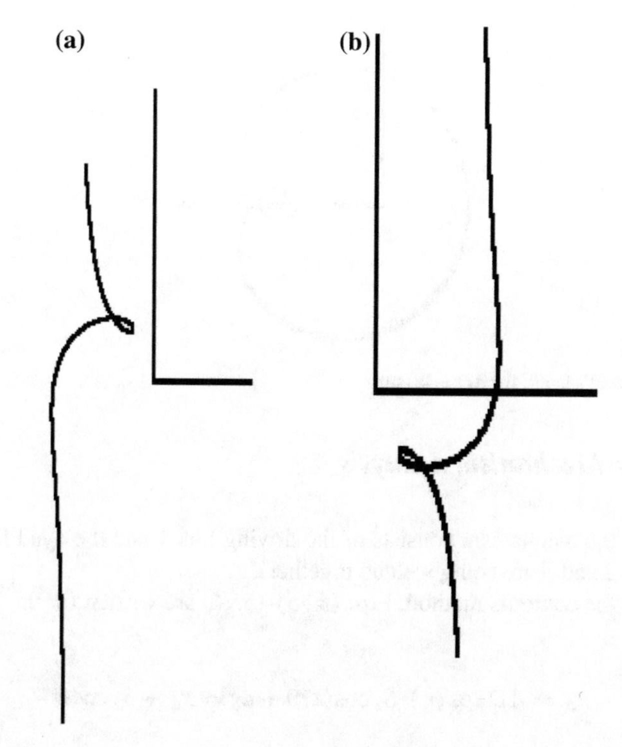

Fig. 5.123 Curve's shape: **a** for $AB = -50$, $ED = 200$ (mm), **b** for $AB = 50$, $ED = -200$ (mm)

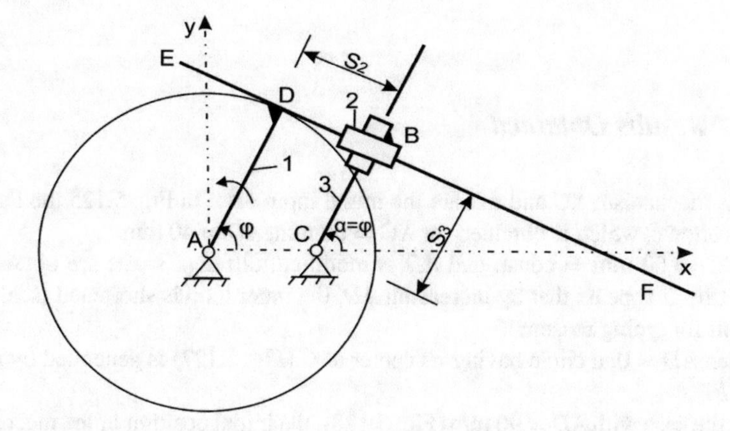

Fig. 5.124 Mechanism obtained

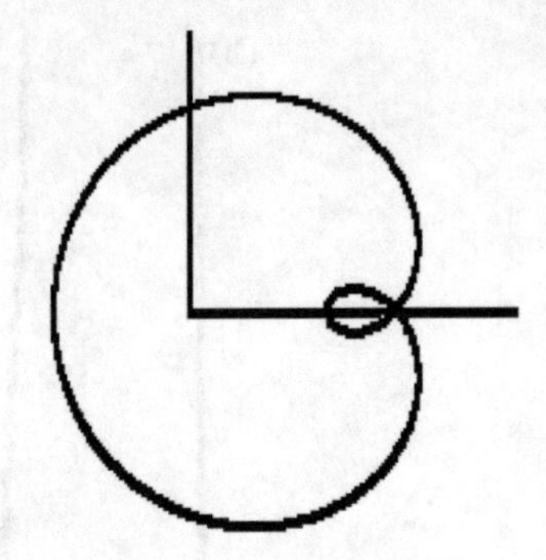

Fig. 5.125 Pascal's snail, for $AD = 40$ mm

5.8.2 The Mechanism Analysis

Structurally, the mechanism consists of the driving link 1 and the dyad BC of PPR type (sliders 2 and 3 are being welded together).

Based on the contours method, Eqs. (5.75)–(5.76) are written for the generating point B:

$$x_B = AD\cos\varphi + S_2\cos(270 + \varphi) = x_C + S_3\cos\varphi \tag{5.75}$$

$$y_B = AD\sin\varphi + S_2\sin(270 + \varphi) = y_C + S_3\sin\varphi \tag{5.76}$$

5.8.3 Results Obtained

For this mechanism XC and AD are the initial input data. In Fig. 5.125 the Pascal's snail is shown, which is obtained for $XC = 60$ mm, $AD = 40$ mm.

If $XC = 60$ mm $=$ const. and AD is modified, different snails are obtained in Fig. 5.126. It appears that by increasing AD, the inner loop is shortened (scales are different for typing reasons).

When $AD = 0$, a circle having its center at C (Fig. 5.127) is generated by points E and F.

For the case with $AD = 90$ mm (Fig. 5.128), the initial position of the mechanism (Fig. 5.129) and the successive positions (Fig. 5.130) were set, observing a continuous motion without jumps.

Further on, it was maintained $AD = 90$ mm $=$ constant and XC was successively modified: $XC = 0$ (Fig. 5.131), $XC = -80$ (mm) (Fig. 5.132). It is noted the Pascal's snail became a circle for $XC = 0$ and for the negative value of XC, it is positioned on the left of Y-axis, oriented opposite to the curve in Fig. 5.126.

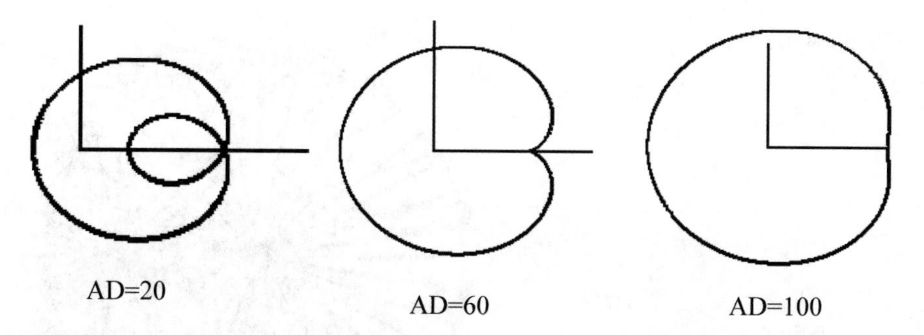

Fig. 5.126 Influence of AD size upon the curve's shape, for $XC = 60$ mm

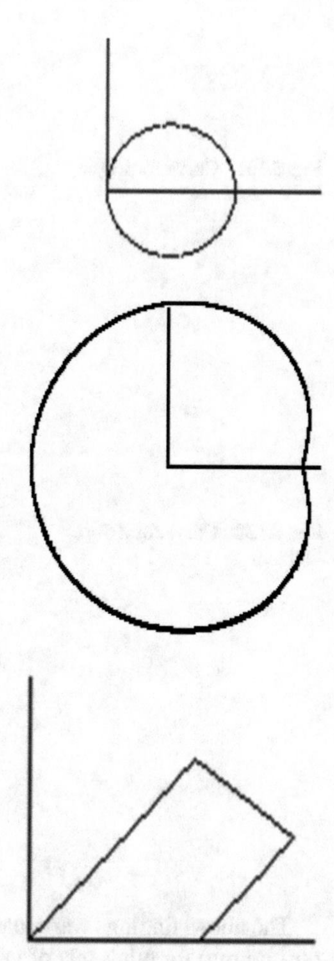

Fig. 5.127 Curve for $AD = 0$ mm

Fig. 5.128 Curve for $AD = 90$ mm

Fig. 5.129 Mechanism in a position

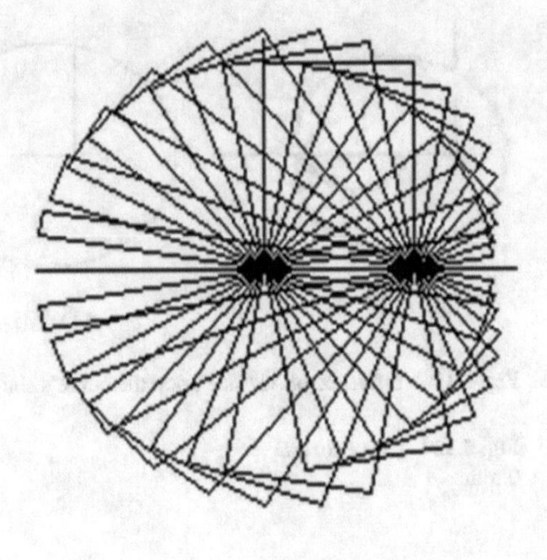

Fig. 5.130 Successive positions

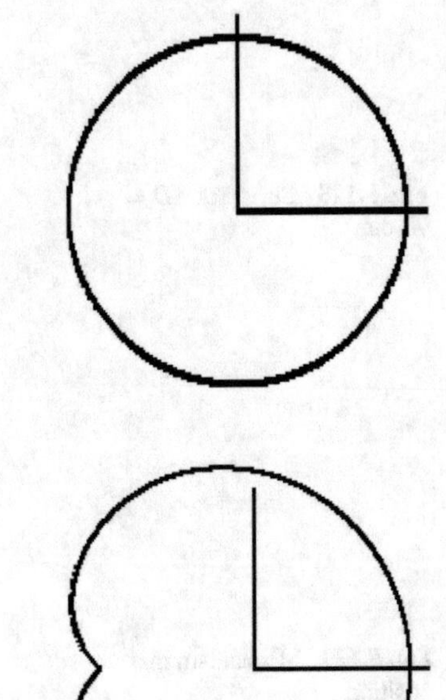

Fig. 5.131 Curve for $XC = 0$

Fig. 5.132 Curve for $XC =$ –80 mm

The above findings were maintained (regarding the curves' dimensions and their positioning) for other sets of values presented in Fig. 5.133:

1. for $XC = 50$ mm, $AD = 11$ mm;
2. for $XC = 50$ mm, $AD = 50$ mm;
3. for $XC = -40$ mm, $AD = 40$ mm.

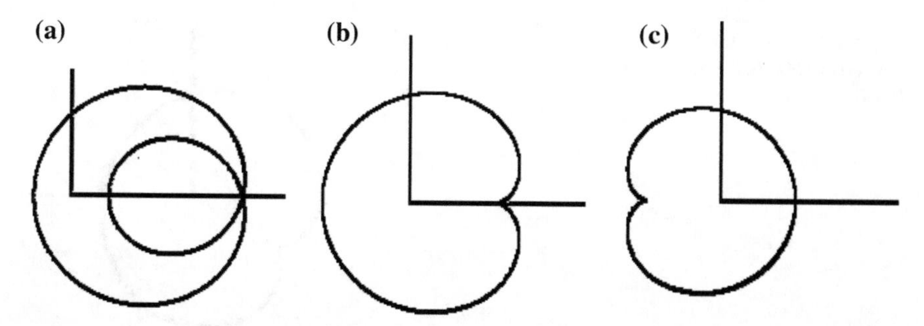

Fig. 5.133 Pascal's snail for other sets of values

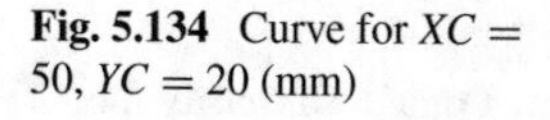

Fig. 5.134 Curve for $XC =$ 50, $YC = 20$ (mm)

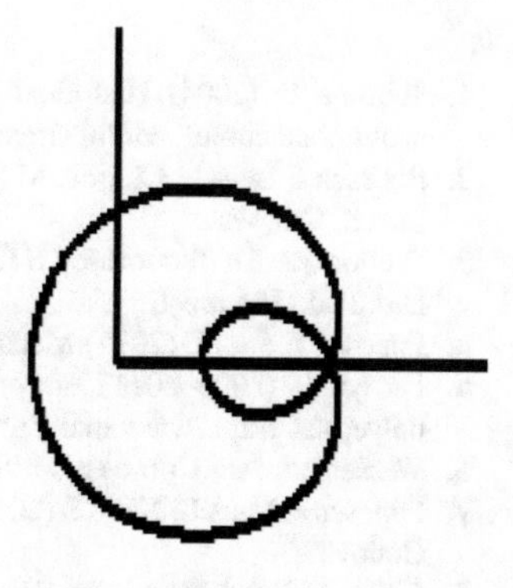

Fig. 5.135 Curve for $YC =$ –20 mm

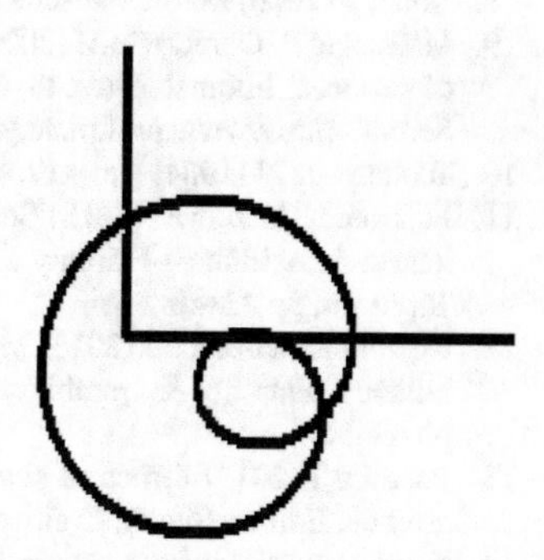

Further on, *AD* was maintained at constant value $AD = 50$ mm, *XC* was changed, but *YC* was introduced, meaning that point *C* is no longer on the *x*-axis. In Fig. 5.134, the case with $XC = 50$ mm, $YC = 20$ mm, *AC* axis being offset with $YC > 0$ is seen, the snail resulting in a counter clockwise rotated position.

With the same data, but with $YC = -20$ mm, the clockwise sloped snail was obtained (Fig. 5.135).

For $XC = -50$ mm, $AD = 50$ mm and $YC = -30$ mm, the snail is oriented with the loop toward the IIIrd quadrant (Fig. 5.136).

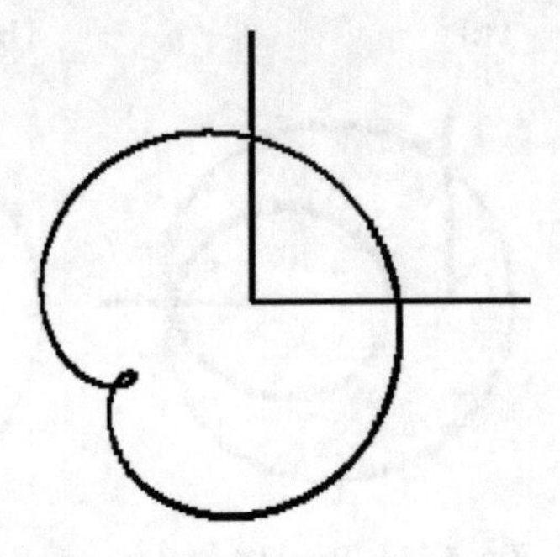

Fig. 5.136 Curve for $XC = -50$, $AD = 50$, $YC = -30$ (mm)

References

1. Taimina D (2004) Historical mechanisms for drawing curves. Cornell University. http://ecommons.cornell.edu/bitstream/1813/2718/1/2004-9.pdf
2. Popescu I, Luca L, Cherciu M (2013) Structura şi cinematica mecanismelor. Aplicaţii. Editura Sitech, Craiova
3. Teodorescu D, Teodorescu STD (1975) Culegere de probleme de geometrie superioară (ed) Did. Ped., Bucureşti
4. Popescu I, Sass L (2001) Mecanisme generatoare de curbe (ed) Scrisul Românesc, Craiova
5. Ferreol R (2006–2011) Encyclopedie des formes remarquables courbes, surfaces, fractals, polyedres. http://www.mathcurve.com/
6. Wassenaar, Jan- Cubic egg. http://www.2dcurves.com/cubic/cubiceg.html
7. Popescu I, Luca L, Mitsi S (2011) Geometria, structura şi cinematica unor mecanisme. Sitech, Craiova
8. Köller J (2000) Mathematische Basteleien. http://www.mathematische-basteleien.de/
9. Malesevic B, Obradovic M (2009) An application of Groebner bases to planarity of intersection of surfaces. Filomat 23(2):43–55 (Faculty of Sciences and Mathematics, University of Nis, Serbia). http://www.pmf.ni.ac.yu/filomat
10. Teodorescu N (1984) ş.a. – Probleme din Gazeta Matematică Editura Tehnica, Bucuresti
11. Popescu I, Cherciu M (2013) Bernoulli quartic generation by an original mechanism. In: Fiabilitate şi durabilitate—Fiability & Durability No. 1/2013. Universitatea «Constantin Brâncuşi», Târgu Jiu, pp 42–48
12. Cherciu M, Popescu I (2013) Mechanism to draw MacLaurin trisectrix. In: Fiabilitate şi durabilitate – Fiability & Durability No. 1/2013. Universitatea «Constantin Brâncuşi», Târgu Jiu, pp 49–55
13. Popescu I (2012) Structura şi cinematica unor mecanisme generatoare de curbe şi suprafeţe estetice. Editura Sitech, Craiova
14. http://www.ajol.info/index.php/ijest/article/viewFile/63714/51541
15. http://ornamentalturning.net/articles/volmer-oval_indexer.pdf
16. http://kmoddl.library.cornell.edu/model.php?m=302

Chapter 6
Mechanism for Generating Spatial Curves

Abstract We designed an algorithm and an original computer program that can draw spatial curves, starting from considerations of spatial geometry and the rules of descriptive geometry. In this way, spatial curves described by points of some elements could be generated. A screw mechanism with four driving links with correlated motions was designed. Curves obtained are cylindrical, conical, and variable pitch propellers. Other points describe spatial curves with a wide variety of shapes. Another original mechanism has a bearing placed on the slider of a connecting-rod mechanism, supporting a rotational motion element, on which there are two other elements with rotational motions (correlated motions). A wide variety of spatial curves were thus obtained.

6.1 Screw Mechanism

6.1.1 Input Data

Based on the geometrical remarks presented in [1, 2, 4, 5], a computer program was developed for spatial curves.

The spatial mechanism in Fig. 6.1 is considered [3]. It receives an external rotational motion with the angle φ, so that the *AFBE* assembly moves up, the screw kinematic pair at point *A*, having the pitch *p*. It is noted that the point *E* is fixed to *FE*, while the point *B* moves on *FE* with the *S* displacement.

The *BD* element rotates about *FB* and the *EC* element rotates about *FE*, while the element *FE* rotates about point *A* and moves along the *z*-axis.

For the mechanism in Fig. 6.1, we can write the Eqs. (6.1–6.15):

$$S = c_1\varphi \tag{6.1}$$

$$\psi = c_2\varphi \tag{6.2}$$

I. Popescu et al., *Mechanisms for Generating Mathematical Curves*,
Springer Tracts in Mechanical Engineering,
https://doi.org/10.1007/978-3-030-42168-7_6

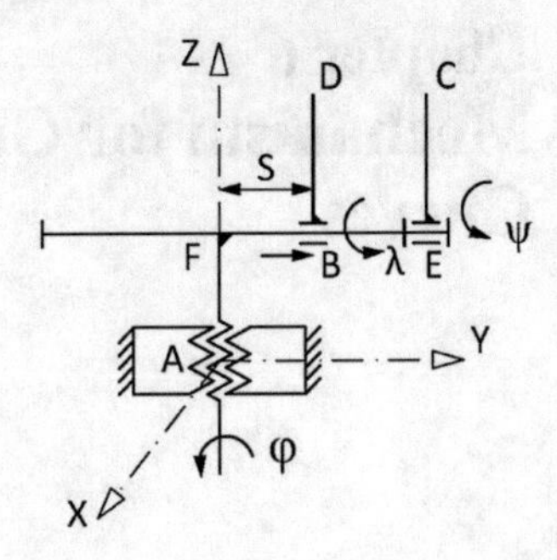

Fig. 6.1 The spatial mechanism

$$\lambda = c_3\varphi \tag{6.3}$$

$$x_B = S\cos\varphi \tag{6.4}$$

$$y_B = S\sin\varphi \tag{6.5}$$

$$z_B = AF + \varphi \cdot p \tag{6.6}$$

$$x_E = FE\cos\varphi \tag{6.7}$$

$$y_E = FE\sin\varphi \tag{6.8}$$

$$z_E = z_B \tag{6.9}$$

$$x_C = x_E + EC\cos\psi \tag{6.10}$$

$$y_C = y_E \tag{6.11}$$

$$z_C = z_E + EC\sin\psi \tag{6.12}$$

$$x_D = x_B + BD\cos\lambda \tag{6.13}$$

$$y_D = y_B \tag{6.14}$$

$$z_D = z_B + BD\sin\lambda \tag{6.15}$$

The range of φ angle is between 0 and 360 for a full rotation, but the element AFE can perform n_1 rotations. The displacement S and the angles ψ and λ are considered linear functions of φ through the coefficients c_1, c_2, and c_3 but may also be nonlinear, in which case it results a greater variety of curves.

To perform the following numerical calculations were adopted: $AF = 0$: $FE = 40$: $EC = 35$: $BD = 45$ [mm].

6.1.2 Generation of Cylindrical Helix

It is considered only the rotational and translational motions of FE, and it generates the trajectory of E point (Fig. 6.1), i.e., a helix of pitch p. We considered a number of n_1 full rotations for AFE.

Some examples of helices obtained for various values of pitch p [mm] and a number of full rotations n_1 are given in Fig. 6.2.

6.1.3 Generation of Conical Helix

If the point B (Fig. 6.1) is moving along FE with the displacement $S = c_1\varphi$, there are obtained different conical helices (Fig. 6.3) as trajectories of E point.

It is found that, changing the direction of displacement S, the curve goes from origin to the right for $S > 0$ and to the left for $S < 0$.

It is possible to obtain the inverted conical helix, that is, to say having the maximum diameter at the base and the minimum at the top, if the point B goes from E to F (Fig. 6.1). Some of the results are given in Fig. 6.4.

In the case of Fig. 6.4 ($p = 5$ mm, $n_1 = 10$, $c_1 = 0.5$), it is noted (from the diagram in Fig. 6.5) that the curve $S(\varphi)$, which is a straight, here is tangential to X_B and Y_B, i.e., when EF is parallel to the y-axis, it is obvious that $S = Y_B$, and when EF is parallel to the x-axis, it results $S = X_B$. The distance S is always decreasing from FE to the minimum. The Z_B curve is linear and always ascendant.

For $p = 5$ mm, $n_1 = 10$ and $c_1 = 1.5$, two successive conical helices, one normal and one larger inverted, are obtained in Fig. 6.6.

For this case, the diagram $S(\varphi)$ results in Fig. 6.7. It is noticed that S decreases from the maximum value which is equal to FE, to zero value, for $\varphi = 1530°$ (abscissa having $\varphi \cdot n_1$ degrees), and after that, the curve of S becomes tangent to the lower limits of the curves X_B and Y_B. The values of S are now negative, the point B is passing to the left of point F. The increases of coordinates X_B and Y_B being in the negative area lead to a conical helix that starts from zero values for X_B and Y_B, and after these grow, the helix being inverted relative to the previous one, but the drawing point rotates contrary to the case of previous helix.

Other examples are given in Fig. 6.8.

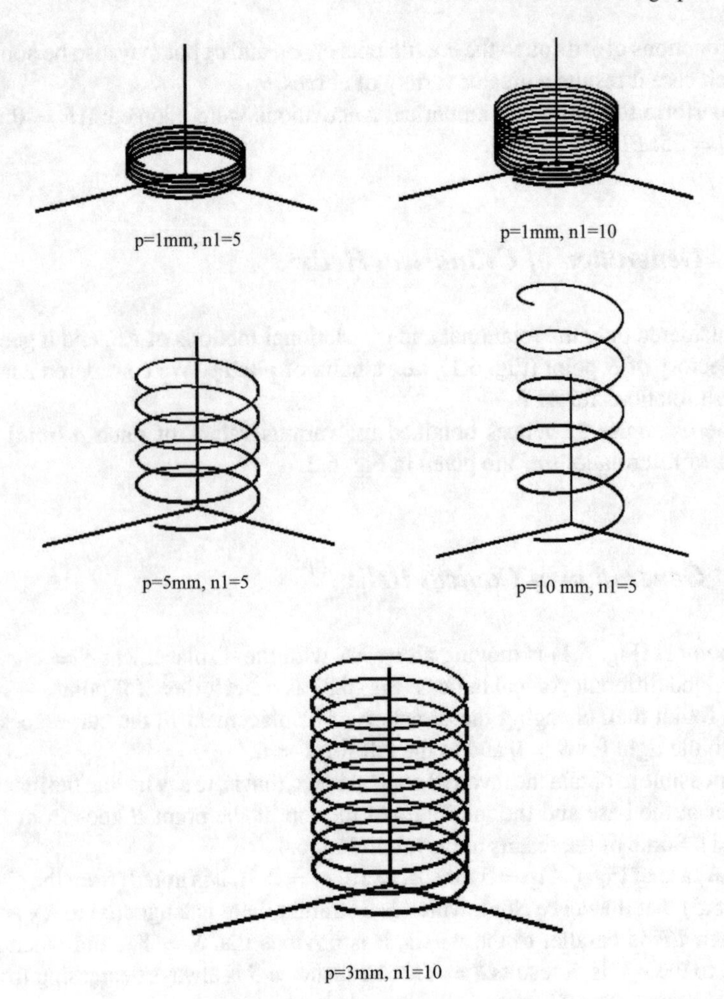

Fig. 6.2 Cylindrical helices obtained

6.1.4 Generation of Complex Spatial Curves

The trajectory of point C

In the case of point C (Fig. 6.1), there are the variables φ and ψ, with $\psi = c_2 \cdot \varphi$. First, specific cases have been taken in order to check shape of the resulted curves. Thus, for $XE = 0$, $YE = EF$ and $ZE = AF = 0$, it was obtained a circle, having FE as axis, which appears in space as an ellipse. For $XE \neq 0$, it results also an ellipse, but in a different position than in the previous case (Fig. 6.9).

Other results are given in Figs. 6.10 and 6.11. There is a wide variety of spatial curves with many branches intersecting.

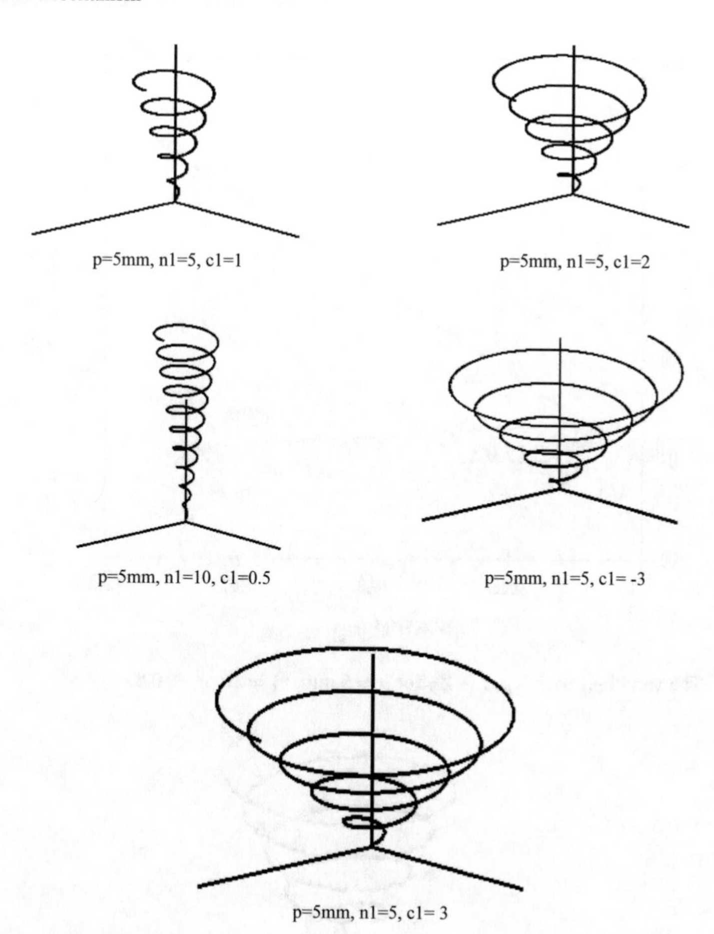

Fig. 6.3 Conical helices obtained

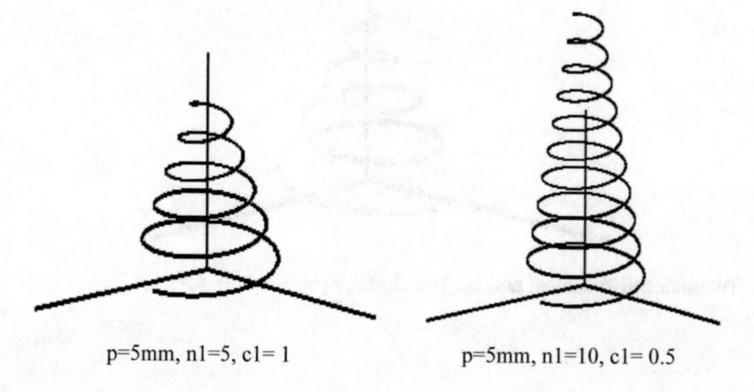

Fig. 6.4 Inverted conical helices

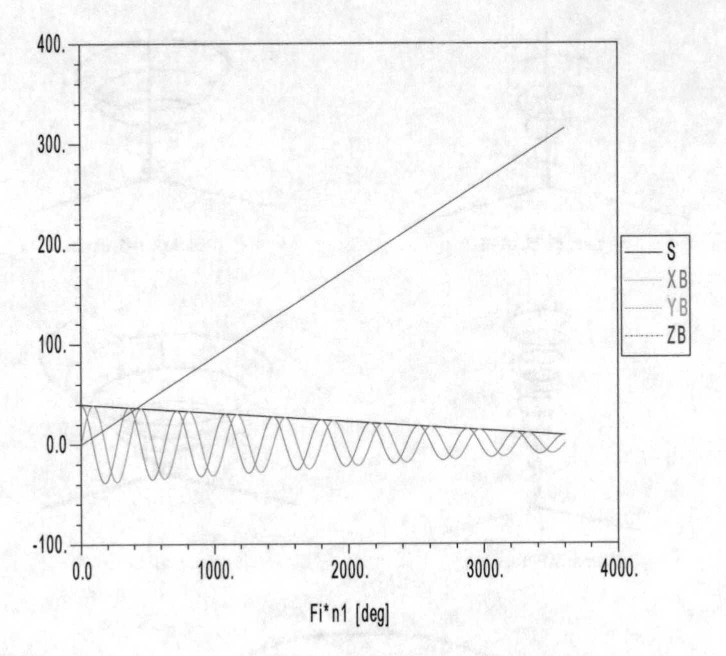

Fig. 6.5 The variations of S, X_B, Y_B, Z_B for $p = 5$ mm, $n_1 = 10$, $c_1 = 0.5$

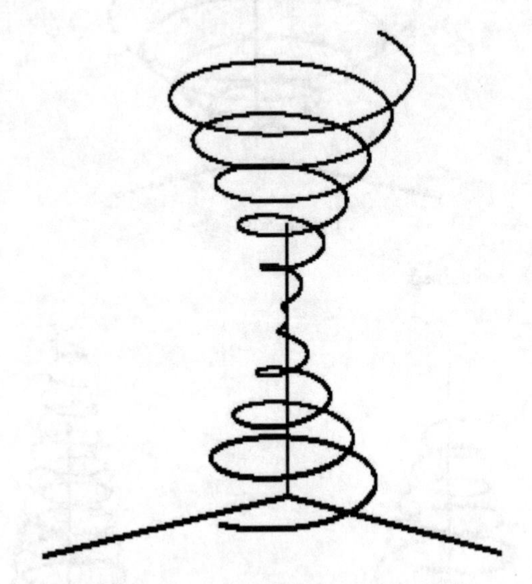

Fig. 6.6 Two successive conical helices ($p = 5$ mm, $n_1 = 10$, $c_1 = 1.5$)

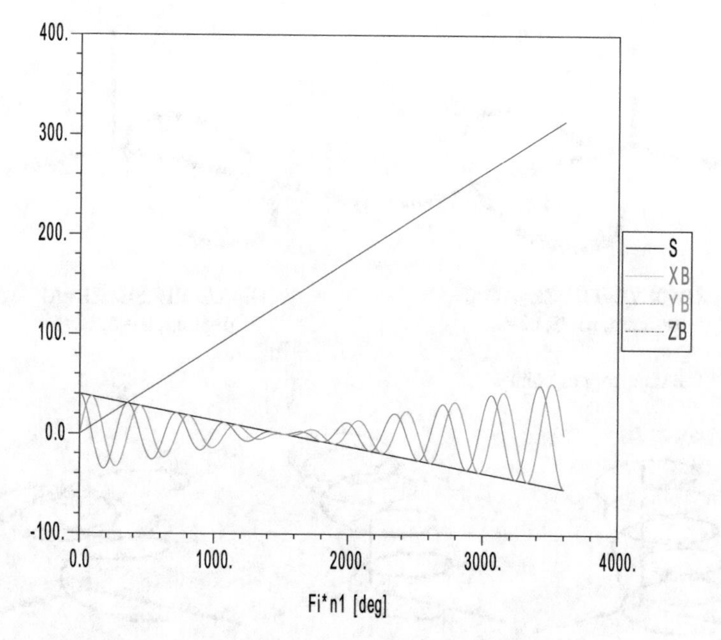

Fig. 6.7 The variations of S, X_B, Y_B, Z_B for $p = 5$ mm, $n_1 = 10$, $c_1 = 1.5$

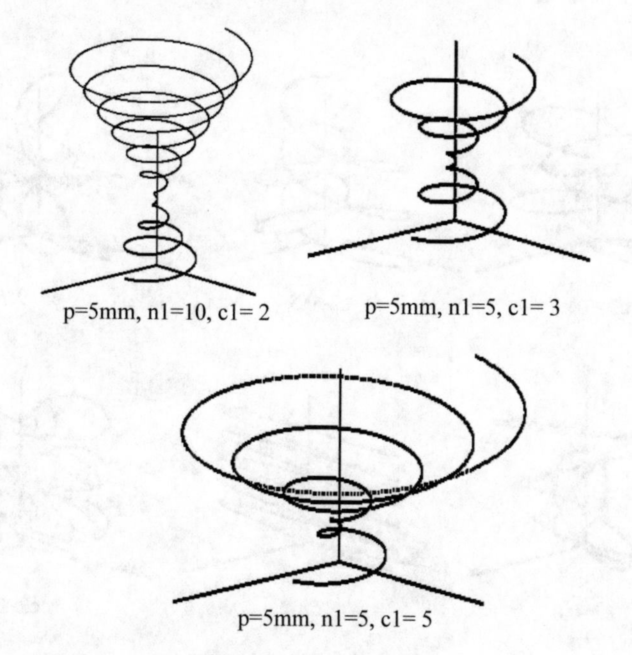

Fig. 6.8 Examples of successive conical helices

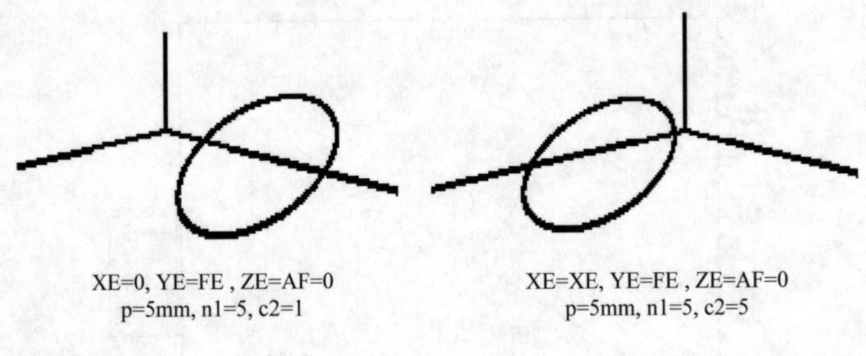

Fig. 6.9 The trajectory of point *C*

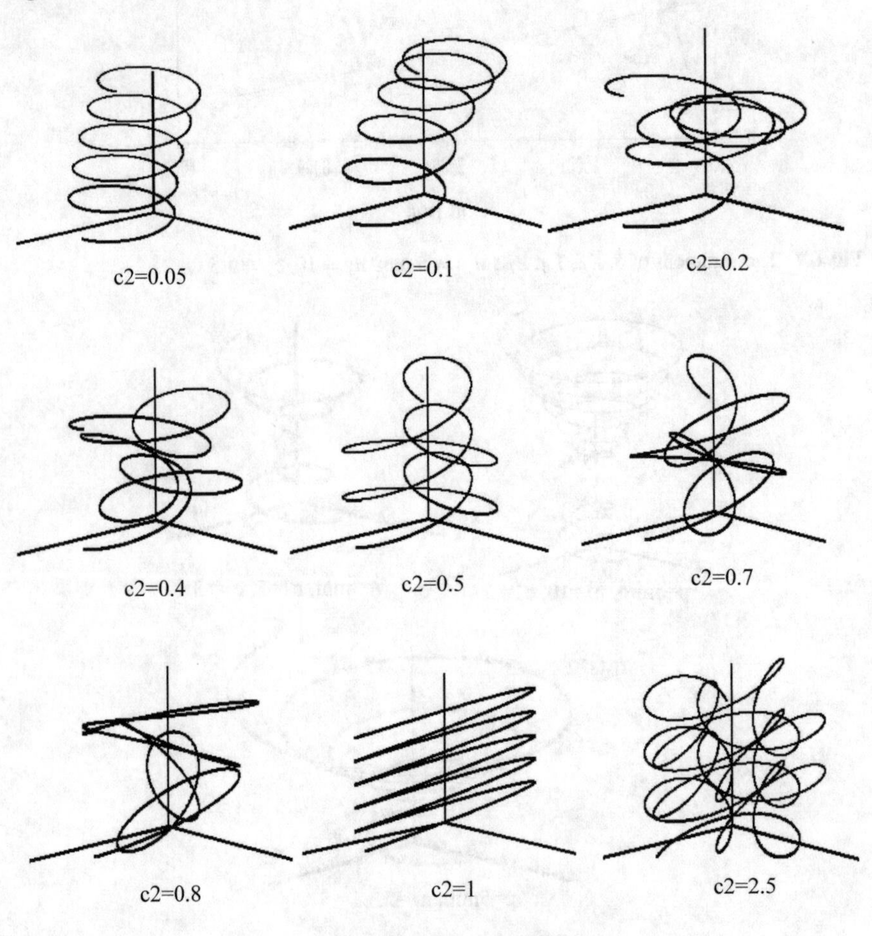

Fig. 6.10 Spatial curves obtained for $p = 5$ mm, $n_1 = 5$, and different c_2

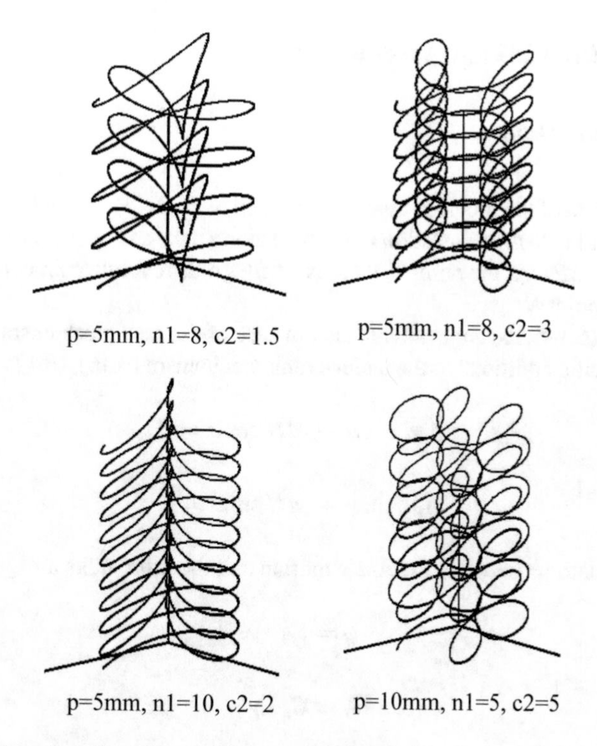

Fig. 6.11 Variety of spatial curves with intersecting branches

The trajectory of point D

The trajectory of point D (Fig. 6.1) depends on φ, S, and λ, with $S = c_1\varphi$ and $\lambda = c_3\varphi$. Some results are given in Fig. 6.12 ($p = 5$ mm, $n_1 = 10$).

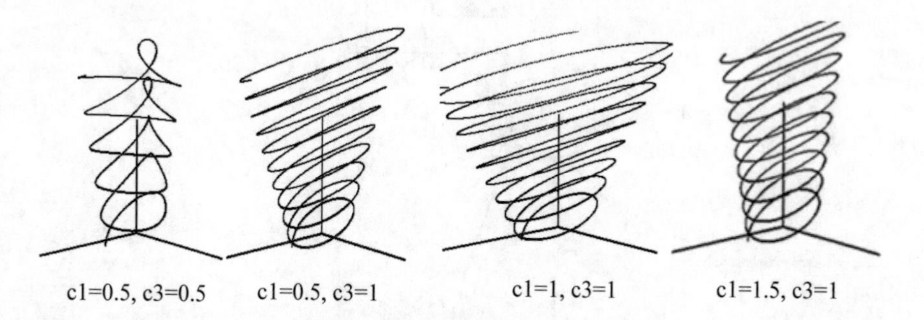

Fig. 6.12 Spatial curves generated as trajectories of point D

6.2 Six-Bar Linkage Mechanism

6.2.1 Input Data

The mechanism of the previous chapter is studied, with the elements *FE*, *EC*, and *BD* (Fig. 6.13) and with the translation movement along the *z*-axis given by a slider-crank mechanism, *AMN*. At the point *T*, there is a fifth grade revolute pair, connected with the slider at point *N*.

The Eqs. (6.1–6.15) are available also in case of the new mechanism (Fig. 6.13), with the specific additions to the slider-crank mechanism (6.16), (6.17):

$$AM\cos\gamma + MN\cos\alpha = 0 \tag{6.16}$$

$$AM\sin\gamma + MN\sin\alpha = h \tag{6.17}$$

The correlations between the *AM*'s motion and the other links are:

$$\varphi = C_4 \cdot \gamma \tag{6.18}$$

$$\lambda = C_3 \cdot \gamma \tag{6.19}$$

$$\psi = C_2 \cdot \gamma \tag{6.20}$$

$$S = C_1 \cdot \gamma \tag{6.21}$$

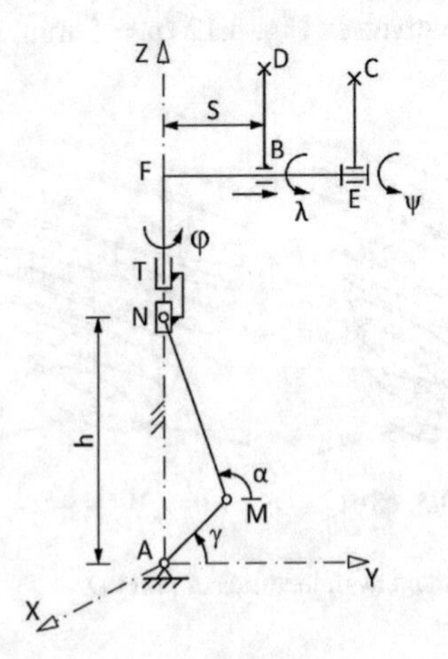

Fig. 6.13 The new mechanism obtained

In this way, it is given a variable displacement on vertical direction, the curves resulted being similar to those generated by the mechanism in Fig. 6.1. The input linkange AM will rotate, but only in the range $(-90 \ldots +90)°$, i.e., only for upward displacement, and not for the downward one, in which case the curves would overlap those of the upward displacement case.

6.2.2 Generation of Cylindrical Helix with Variable Pitch

The trajectory of point E

The trajectory of point E (Fig. 6.13) is a cylindrical helix. Examples are given in Fig. 6.14 to observe the helix variable pitch. The coefficient c_4 which defines φ (6.18) was modified.

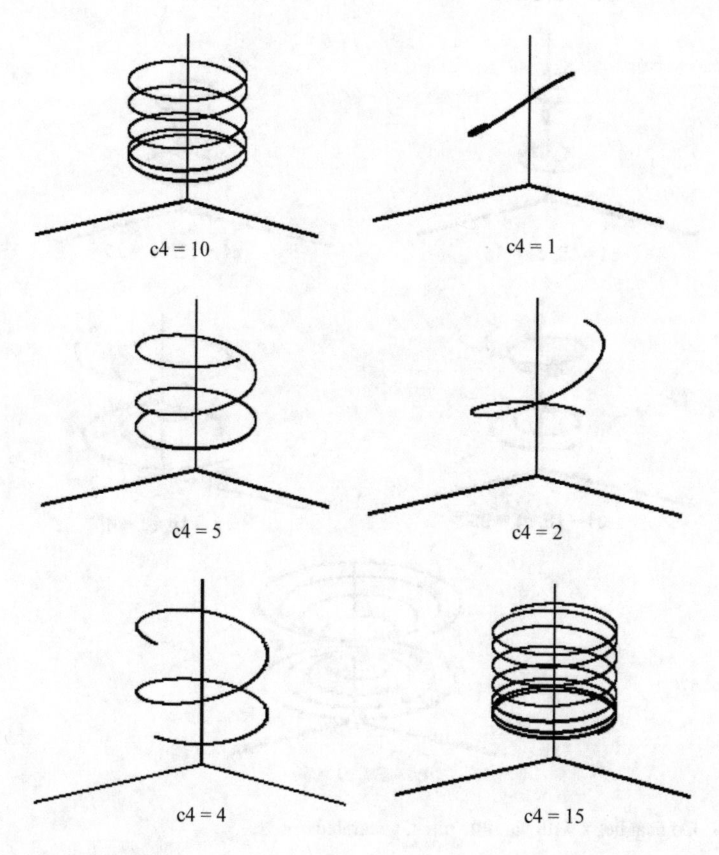

Fig.6.14 Cylindrical helices with variable pitch, generated

6.2.3 Generation of Conical Helix with Variable Pitch

In this case, for the small values of c_1 (6.21), the conical helix has small dimensions, and at high values, the point B moves to the left of point F, i.e., it has negative values. In this way, the trajectory of point E gives two successive conical helices traced in opposite directions (Fig. 6.15).

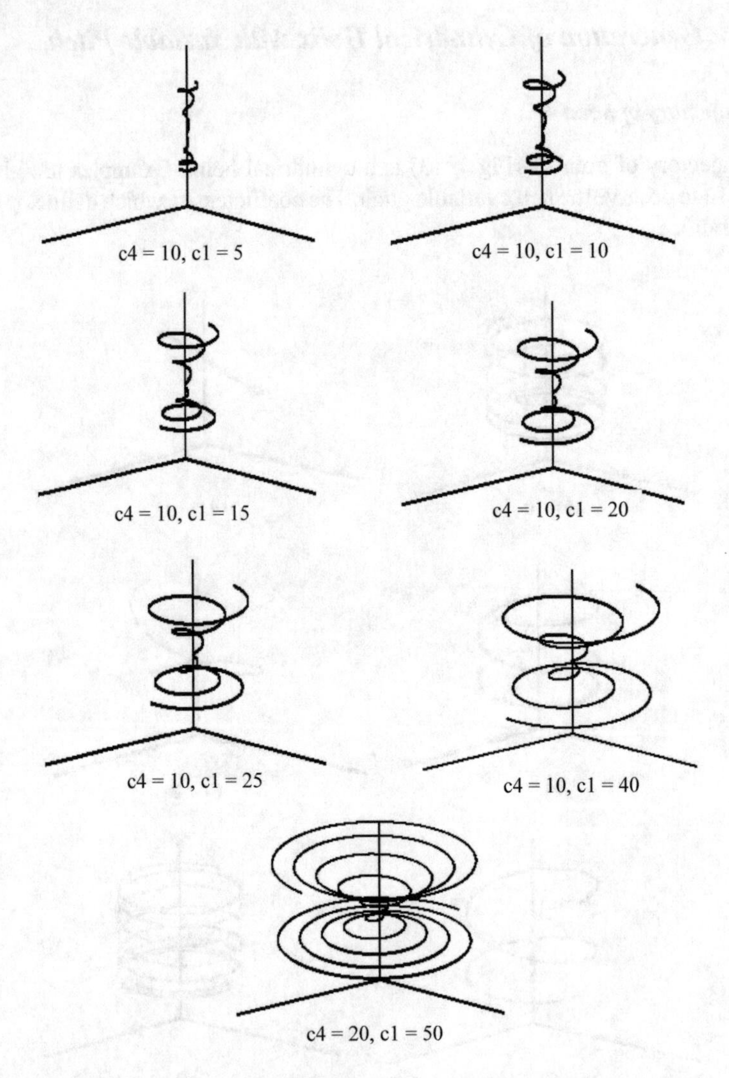

Fig. 6.15 Conical helix with variable pitch, generated

6.2.4 Generation of Complex Spatial Curves with Variable Pitch

The trajectory of point C

In the case that elements *AM*, *FE*, and *EC* (Fig. 6.13) rotate around different axes, the trajectory of point *C* is a complex spatial curve, noting however a variable pitch, given by the slider-crank mechanism (Fig. 6.16).

The trajectory of point D

For the mechanism in Fig. 6.13, the trajectory of point *D* generates spatial complicated curves, but at which it is noticeable a variable pitch (Fig. 6.17).

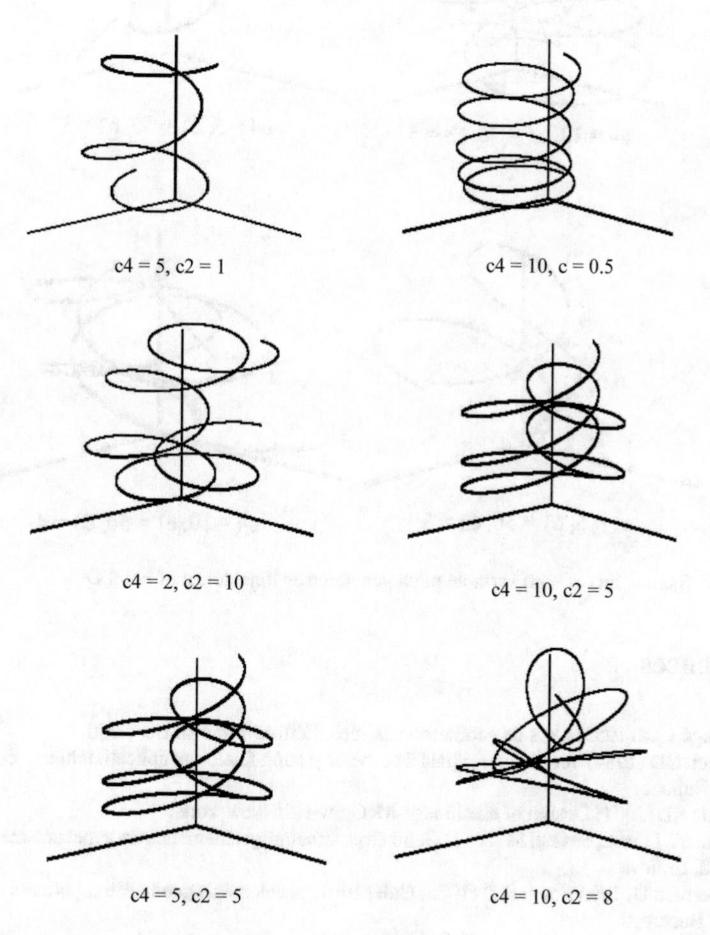

Fig. 6.16 Complex spatial curves with variable pitch, generated

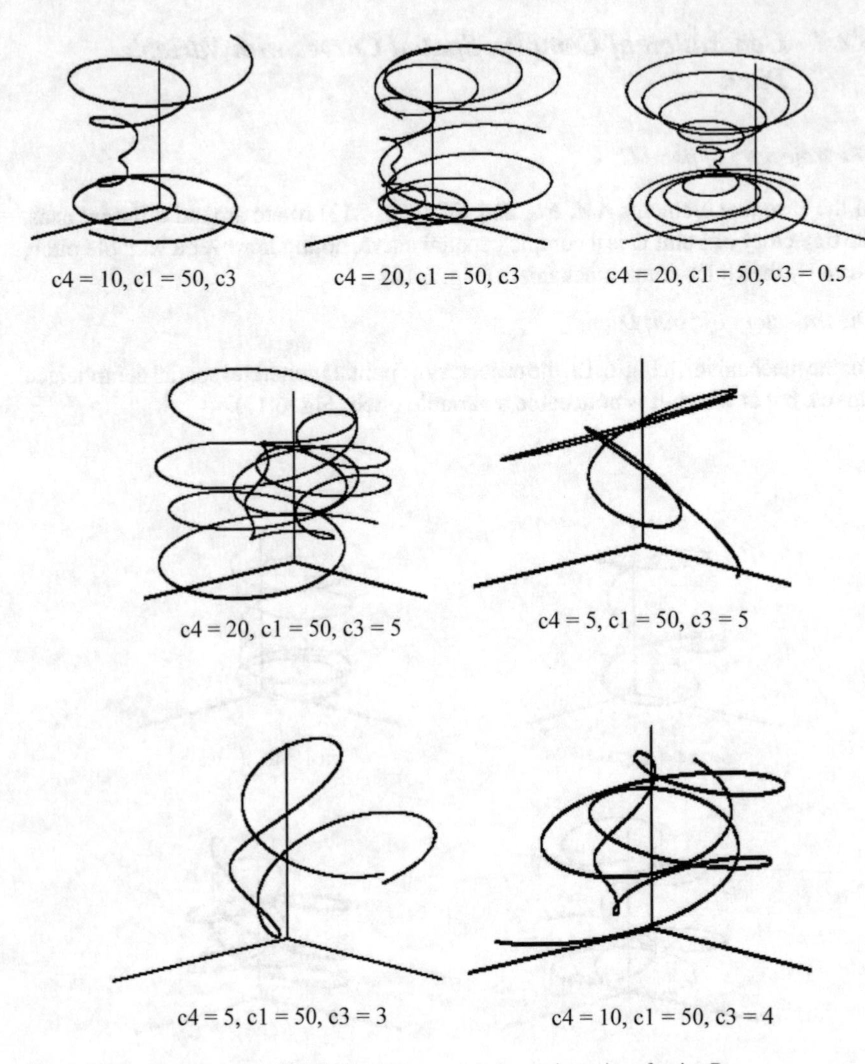

Fig. 6.17 Spatial curves with variable pitch generated as trajectories of point *D*

References

1. Creangă I ş.a (1953) Curs de geometrie analitică. Editura Tehnică, Bucureşti
2. Ionescu GD (1984) Teoria diferenţială a curbelor şi suprafeţelor cu aplicaţii tehnice. Ed. Dacia, Cluj-Napoca
3. Norton RL (2004) Design of machinery. McGraw-Hill, New York
4. Popescu I, Luca L, Sevasti M (2011) Geometria, structura şi cinematica unor mecanisme. Editura Sitech, Craiova
5. Teodorescu D, Teodorescu ŞtD (1975) Culegere de probleme de geometrie superioară. Ed. Did. Ped., Bucureşti